T0195523

Evaluation of Options for Overseas Combat Support Basing

Mahyar A. Amouzegar, Ronald G. McGarvey, Robert S. Tripp, Louis Luangkesorn, Thomas Lang, Charles Robert Roll, Jr.

Prepared for the United States Air Force

PROJECT AIR FORCE

The research described in this report was sponsored by the United States Air Force under Contract F49642-01-C-0003. Further information may be obtained from the Strategic Planning Division, Directorate of Plans, Hq USAF.

Library of Congress Cataloging-in-Publication Data

Evaluation of options for overseas combat support basing / Mahyar A. Amouzegar ... [et al.].
     p. cm.
   "MG-421."
   ISBN 0-8330-3874-5 (pbk. : alk. paper)
     1. Air bases, American. 2. Airlift, Military—United States. 3. United States— Armed Forces—Supplies and stores. 4. United States. Air Force—Foreign service. I. Amouzegar, Mahyar A.

UG634.49.E88 2006
358.4'1621—dc22

                                                                    2005030192

The RAND Corporation is a nonprofit research organization providing objective analysis and effective solutions that address the challenges facing the public and private sectors around the world. RAND's publications do not necessarily reflect the opinions of its research clients and sponsors.

RAND® is a registered trademark.

Cover Photograph by Staff Sergeant Jennie Ivey, Courtesy U.S. Air Force

Published 2006 by the RAND Corporation
1776 Main Street, P.O. Box 2138, Santa Monica, CA 90407-2138
1200 South Hayes Street, Arlington, VA 22202-5050
4570 Fifth Avenue, Suite 600, Pittsburgh, PA 15213
RAND URL: http://www.rand.org/
To order RAND documents or to obtain additional information, contact Distribution Services: Telephone: (310) 451-7002;
Fax: (310) 451-6915; Email: order@rand.org

Preface

This work evaluates overseas combat support basing options for storing war reserve materiel (WRM). These option packages, or "portfolios," have differing numbers and types of forward support locations (FSLs), e.g., land-based or afloat, and differing allocations of WRM at alternative sites. The evaluations of these packages address the effectiveness and efficiency of the options in meeting a wide variety of potential scenarios. In this work, we have developed capability-based analytic tools to evaluate the tradeoffs among various options. A central element of our analytic framework is an optimization model that allows us to select the "best" mix of land- and sea-based FSLs for a given operational scenario based on several criteria.

During the past eight years, the RAND Corporation has studied options for configuring an Agile Combat Support (ACS) system that would enable the Air and Space Expeditionary Force (AEF) goals of rapid deployment, immediate employment, and uninterrupted sustainment from a force structure located primarily within the continental United States (CONUS). This report is one of a series of RAND publications that address ACS options.

This research, conducted in the Resource Management Program of RAND Project AIR FORCE, was sponsored by the Air Force Deputy Chief of Staff for Installations and Logistics (AF/IL).

The report should be of interest to logisticians, operators, and mobility planners throughout the Department of Defense (DoD), especially those in the Air Force. Other publications issued as part of the Supporting Air and Space Expeditionary Forces series include the following:

- *An Integrated Strategic Agile Combat Support Planning Framework*, Robert S. Tripp, Lionel A. Galway, Paul S. Killingsworth, Eric Peltz, Timothy L. Ramey, and John G. Drew (MR-1056-AF, 1999). This report describes an integrated combat support (CS) planning framework that may be used to evaluate support options on a continuing basis, particularly as technology, force structure, and threats change.
- *New Agile Combat Support Postures*, Lionel A. Galway, Robert S. Tripp, Timothy L. Ramey, and John G. Drew (MR-1075-AF, 2000). This report describes how alternative resourcing of forward operating locations (FOLs) can support employment timelines for future AEF operations. It finds that rapid employment for combat requires some prepositioning of resources at FOLs.
- *An Analysis of F-15 Avionics Options*, Eric Peltz, Hyman L. Shulman, Robert S. Tripp, Timothy L. Ramey, and John G. Drew (MR-1174-AF, 2000). This report examines alternatives for meeting F-15 avionics maintenance requirements across a range of likely scenarios. The authors evaluate investments for new F-15 Avionics Intermediate Shop test equipment against several support options, including deploying maintenance capabilities with units, performing maintenance at FSLs, or performing all maintenance at the home station for deploying units.
- *A Concept for Evolving to the Agile Combat Support/Mobility System of the Future*, Robert S. Tripp, Lionel A. Galway, Timothy L. Ramey, Mahyar A. Amouzegar, and Eric Peltz (MR-1179-AF, 2000). This report describes the vision for the ACS system of the future based on individual commodity study results.
- *Expanded Analysis of LANTIRN Options*, Amatzia Feinberg, Hyman L. Shulman, Louis W. Miller, and Robert S. Tripp (MR-1225-AF, 2001). This report examines alternatives for meeting Low Altitude Navigation and Targeting Infrared for

Night (LANTIRN) support requirements for AEF operations. The authors evaluate investments for new LANTIRN test equipment against several support options, including deploying maintenance capabilities with units, performing maintenance at FSLs, or performing all maintenance at Continental United States (CONUS) support hubs for deploying units.

- *Alternatives for Jet Engine Intermediate Maintenance*, Mahyar A. Amouzegar, Lionel A. Galway, and Amanda Geller (MR-1431-AF, 2002). This report evaluates the manner in which Jet Engine Intermediate Maintenance (JEIM) shops can best be configured to facilitate overseas deployments. The authors examine a number of JEIM support options, which are distinguished primarily by the degree to which JEIM support is centralized or decentralized. See also *Engine Maintenance Systems Evaluation (En Masse): A User's Guide*, Mahyar A. Amouzegar and Lionel A. Galway (MR-1614-AF, 2003).

- *A Combat Support Command and Control Architecture for Supporting the Expeditionary Aerospace Force*, James Leftwich, Robert S. Tripp, Amanda Geller, Patrick H. Mills, Tom LaTourrette, C. Robert Roll, Jr., Cauley Von Hoffman, and David Johansen (MR-1536-AF, 2002). This report outlines the framework for evaluating options for combat support execution planning and control. The analysis describes the combat support command and control operational architecture as it is now and as it should be in the future. It also describes the changes that must take place to achieve that future state.

- *Reconfiguring Footprint to Speed Expeditionary Aerospace Forces Deployment,* Lionel A. Galway, Mahyar A. Amouzegar, Richard J. Hillestad, and Don Snyder (MR-1625-AF, 2002). This report develops an analysis framework—footprint configuration—to assist in devising and evaluating strategies for footprint reduction. The authors attempt to define footprint and to establish a way to monitor its reduction.

- *Analysis of Maintenance Forward Support Location Operations,* Amanda Geller, David George, Robert S. Tripp, Mahyar A. Amouzegar, and C. Robert Roll, Jr. (MG-151-AF, 2004). This report discusses the conceptual development and recent implementation of maintenance forward support locations (also known as centralized intermediate repair facilities [CIRFs]) for the United States Air Force. The analysis focuses on the years leading up to and including the AF/IL CIRF test, which tested the operations of centralized intermediate repair facilities in the European theater from September 2001 to February 2002.

- *Lessons from Operation Enduring Freedom*, Robert S. Tripp, Kristin F. Lynch, John G. Drew, and Edward W. Chan, (MR-1819-AF, 2004). This report describes the expeditionary ACS experiences during the war in Afghanistan and compares them with those associated with Joint Task Force Noble Anvil (JTF-NA), the air war over Serbia. This report analyzes how ACS concepts were implemented, compares current experiences to determine similarities and unique practices, and indicates how well the ACS framework performed during these contingency operations. From this analysis, the ACS framework may be updated to better support the AEF concept.

- *A Methodology for Determining Air Force Deployment Requirements*, Don Snyder and Patrick H. Mills (MG-176-AF, 2004). This report outlines a methodology for determining manpower and equipment deployment requirements. It describes a prototype policy analysis support tool based on this methodology, the Strategic Tool for the Analysis of Required Transportation (START), and generates a list of capability units, called Unit Type Codes (UTCs), that are required to support a user-specified operation. The program also determines movement characteristics. A fully implemented tool based on this prototype should prove to be useful to the Air Force in both deliberate and crisis action planning.

- *Lessons from Operation Iraqi Freedom*, Kristin F. Lynch, John G. Drew, Robert Tripp, and C. Robert Roll, Jr. (MG-193-AF, 2005). This report describes the expeditionary ACS experiences during the war in Iraq and compares them with those associated with Joint Task Force Noble Anvil (JTF-NA), in Serbia and Operation Enduring Freedom, in Afghanistan. This report analyzes how combat support performed and how ACS concepts were implemented in Iraq, compares current experiences to determine similarities and unique practices, and indicates how well the ACS framework performed during these contingency operations.

• *Analysis of Combat Support Basing Options,* Mahyar A. Amouzegar, Robert S. Tripp, Ronald G. McGarvey, Edward W. Chan, and C. Robert Roll, Jr. (MG-261-AF, 2004). This report evaluates a set of global FSL basing and transportation options for storing war reserve materiel. The authors present an analytical framework that can be used to evaluate alternative FSL options. A central component of the authors' framework is an optimization model that allows a user to select the best mix of land-based and sea-based FSLs for a given set of operational scenarios, thereby reducing costs while supporting a range of contingency operations.

RAND Project AIR FORCE

RAND Project AIR FORCE (PAF), a division of the RAND Corporation, is the U.S. Air Force's federally funded research and development center for studies and analyses. PAF provides the Air Force with independent analyses of policy alternatives affecting the development, employment, combat readiness, and support of current and future aerospace forces. Research is conducted in four programs: Aerospace Force Development; Manpower, Personnel, and Training; Resource Management; and Strategy and Doctrine.

Additional information about PAF is available on our Web site at http://www.rand.org/paf.

Dedication

This book is dedicated to Hy Shulman (1919–2005), a great friend and mentor who always sought to do the right thing.

Contents

Figures

Tables

Summary

Background

The geopolitical divide that once defined the U.S. military policy collapsed as the Soviet Union disintegrated and was replaced by the rise of regional hegemons, producing an evolving security environment that is driven not only by regional powers but also by a persistent global insurgency and counterinsurgency. The ability of U.S. forces to provide swift and tailored responses to a multitude of threats across the globe is a crucial component of security in today's complex political environment. The Air Force, like the other services, has responded by transforming itself into a more expeditionary force. In order for the Air Force to realize its goals of global strike and persistent dominance, it is vital that the Air Force support the warfighter seamlessly and efficiently in all phases of deployment, employment, and redeployment. One of the major pillars for achieving these objectives is a global combat support basing architecture.

This report focuses on an analytic framework for evaluating options for overseas combat support basing (or forward support locations). The presentation of this framework is important because it addresses how to assess these options in terms of the relevant programming costs while considering a novel approach to scenario planning. This formulation minimizes the costs of operating and constructing facilities and transporting WRM, costs that are associated with meeting the training and deterrent exercises needed to demonstrate U.S. global power projection capability and thereby deter aggression, while maintaining the necessary storage capacity and system throughput to engage in major combat operations should deterrence fail.

This framework is based on the notion that U.S. interests are not only global but dynamic as well, particularly when the United States is confronted with emerging anti-access and area denial threats. Consequently, the U.S. Air Force must be ready to deploy forces quickly across a wide range of potential scenarios.

The Tenets of Deployment Scenarios

As recently as a few years ago, the focus of contingency planners was on individual deliberate threat-based deployments. This led to supporting the warfighter by developing *optimal* combat support networks, which were designed to support known threats. An unfortunate characteristic of this type of designed network is that it often performs poorly if the set of demands (locations and quantities) differs from the plan. The new planning environment, with its broad (and unclear) set of potential adversaries, calls for *robust* and *efficient* combat support networks that, while not necessarily optimal for any one deliberate plan, meet operational requirements at reasonable costs over a wide range of contingencies. We have developed a new framework that integrates the traditional threat-based assessments concept with capability-based planning. This framework relies on a sequenced, potentially simultaneous set of deployment scenarios, which we call the Multi-Period–Multi-Scenario (MPMS) concept.

In keeping with this security paradigm and the concept of MPMS, we constructed a deployment framework using the following tenets:

- The combat support basing architecture should be developed using a global perspective and not centered on a few disconnected areas.

- A wide range of plausible deployment scenarios should be considered.
- Deployments should be sequenced in time and space.
- Different sets of deployment scenarios or "streams of reality" should be used to hedge against uncertainty.

Analysis Approach

To evaluate and select alternative forward basing options, we developed an analytic framework that uses an optimization model to assess the cost and capability of various portfolios of overseas combat support basing or forward support locations (FSLs) for meeting a wide variety of global force projections.

We have taken two complementary approaches in developing the optimization model: The primary approach attempts to minimize the overall system cost while meeting operational requirements; the other approach focuses on maximizing the support capability (e.g., reducing the time to initial operating capability). Examining the costs of alternative support basing options, for a constant level of performance against a variety of deployments, is an important process in the development of suitable programming and budgeting plans. In this approach, we are careful to ensure that adequate capacity is maintained to meet requirements as specified in the Defense Planning Scenarios.

Our analyses show the costs and deployment timelines for various FSL options under different degrees of stress on combat support while taking into account infrastructure richness, basing characteristics, deployment distances, strategic warning, transportation constraints, dynamic requirements, and reconstitution conditions. We developed several sets of deployment scenarios using the MPMS concept, with each including training exercises, deterrent missions, and major combat operations. These so-called "streams of reality" allow our model to measure the effect of timing, location, and intensity of operational requirements on combat support—and vice versa. We develop several of these streams (or *timelines*) to account for the inherent uncertainties in future planning associated with each timeline.

After we determine the desired requirements in terms of combat support resources, our optimization model, the *RAND Overseas Basing Optimization Tool (ROBOT)*, selects a set of FSL locations that would minimize the costs of supporting these various deterrence and training exercises while maintaining the capability to support major regional conflicts should deterrence fail. This tool essentially allows for the analysis of various "what-if" questions and assesses the solution set in terms of resource costs for differing levels of combat support capability.

Our analytic approach has several steps (see Figure S.1):

1. We first select a diverse set of deployment scenarios that would stress the combat support system. These deployments include small-scale humanitarian operations, continuous force presentation to deter aggression, and major combat operations.
2. The deployments and the force options drive the requirements for combat support, such as base operating support equipment, vehicles, and munitions.

Figure S.1
Overview of the Analytic Process for the Optimization Model

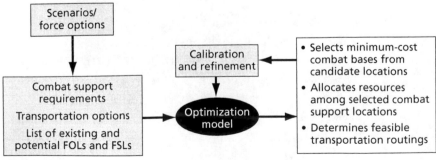

RAND *MG421-S.1*

3. These requirements, the set of potential FSLs and forward operating locations (FOLs), and the transportation options (e.g.,
allowing sealift or not) serve as the inputs to the optimization
model.

4. The optimization model selects the FSL locations that minimize
the costs of operating and constructing facilities and transporting
WRM—costs associated with planned operations, training missions, and deterrent exercises that are scheduled to take place over
an extended time horizon, satisfying time-phased demands for
combat support commodities at FOLs. Major combat operations
are included in this analysis to ensure that the resulting network
has sufficient capability to allow for such operations should deterrence fail; however, the transportation costs associated with these
operations are not considered in the model because of the different funding mechanisms for the execution of combat operations.
The model also optimally allocates the programmed resources and
commodities to those FSLs. It computes the type and the number
of transportation vehicles required to move the materiel to the
FOLs. The result is the creation of a robust transportation and
allocation network that connects a set of disjointed FSL and FOL
nodes.

5. The final step in our approach is to refine and recalibrate the solution set by applying political, geographical, and vulnerability constraints based on current expert judgments concerning the global
environment. Because this step is applied postoptimally and may
make additional iterations necessary, it enables reevalution and re-
assessment of the parameters and options chosen.

The end result of this analysis is a portfolio containing alternative sets of FSL postures, including allocations of WRM to the FSLs,
which can then be presented to decisionmakers. This portfolio will
allow policymakers to assess the merits of various options from a
global perspective.

Combat Support Factors

Several major constraining and contributing factors affect the capability of FSLs to support the warfighter. Our analytic framework takes each of these parameters into account in its process of selecting an optimal set of combat support locations.

Base Access

This important issue deserves careful consideration and must be addressed before each conflict or operation. However, rather than eliminating some sites a priori because of potential political access problems, we allowed the model to select the most desirable sites based on other factors first. We then "forced" specific sites out of the solution set if we had reason to believe that these sites presented access issues—thereby providing the economic cost of restricting the solution to politically acceptable sites.

Forward Support Location Capability and Capacity

The parking space, the runway length and width, the fueling capability, and the capacity to load and offload equipment are all important factors in selecting an airfield to support an expeditionary operation.[1] Runway length and width are key planning factors and are commonly used as first criteria in assessing whether an airfield can be selected.

Airlift and Airfield Throughput Capacity

Timely delivery of combat support materiel is essential in an expeditionary operation. However, a mere increase in the aircraft fleet size may not improve the deployment timelines. The fleet size must always be determined with respect to the throughput capacity of an airfield. The maximum-on-ground (MOG) capability, for example, directly contributes to the diminishing return of deployment time as a function of available airlift.

---

[1] In our analysis, some of these factors are computed parametrically in order to assess a minimum requirement of a potential field for meeting a certain capability.

Forward Operating Location Distance

Distance from FSLs to FOLs can impede expeditionary operations. As the number of airlift aircraft increases, the difference in deployment time caused by distance becomes less pronounced. Adding more airlifters to the system will reduce the deployment time, albeit at a diminishing rate, until the deployment time levels off as a result of MOG constraints.

Modes of Transportation

There are several advantages to using sealift or ground transportation in place of, or in addition to, airlift. Allowing for alterative modes of transportation might bring some FSLs into the solution set that otherwise may have been deemed infeasible or too costly. Ships have a higher hauling capacity than do aircraft and can easily carry outsized or super-heavy equipment. In addition, ships do not require overflight rights from any foreign government.

Afloat Prepositioning

We examined the potential for storing combat support resources (munitions and nonmunitions) aboard an afloat preposition fleet (APF). Although afloat prepositioning does offer additional flexibility and reduced vulnerability versus land-based storage, the APF is much more expensive than land-based storage and presents a serious risk with regard to deployment time. Even if a generous advance warning is assumed to allow for steaming toward a scenario's geographic region, it can be difficult to find a port that is capable of handling these large cargo ships. The requirements placed on the port, including preemption of other cargo movement, also restrict the available ports that can be used by an APF.

Cost

The main objective of the model is to reduce the total cost of exercises and deterrent missions while meeting the time-phased operational demand for combat support resources (for those missions as well as for major combat operations). These costs include construction and/or expansion of facilities, operations and maintenance

(O&M), and transportation for peacetime and training missions. Incorporated in each of these costs is the effect of differences in regional cost-of-living or country cost factors.

Results

We focused on three of the most important combat support resources: Basic Expeditionary Airfield Resources (BEAR), munitions, and rolling stock (e.g., trucks).[2] These resources comprise the bulk of many of the consumable and repairable items in the combat support package; and, in the case of munitions, they pose storage and transport complexities.

From the outset of the study, we attempted to answer two basic questions: How capable are the Air Force's current overseas combat support bases of managing the future environment? And what are the costs and benefits of using additional or alternative overseas combat support bases for storing heavy combat support materiel?

To answer these questions, we devised five different *streams of reality*—or deployment timelines—to represent a wide range of possible future Air Force deployments across the globe (see Table S.1).

The base scenario, or the "most likely global deterrent scenario," places the focus on supporting a number of deployments in the Persian Gulf region, Asian littoral, and North Africa over a time horizon of six years, in keeping with the Future Years Defense Program (FYDP) convention. Figure S.2 represents the size, in terms of combat support requirements, and the timing of each deployment for the base scenario. The sizes of recent deployments are given on the y-axis as a reference. Notice that we have "scheduled" the MCOs in each scenario for execution at the end of the FYDP period. This approach

---

[2] BEAR provides the required airfield operational capability (such as housekeeping or industrial operations) to open an austere or semi-austere airbase.

Table S.1
Sequencing of Scenarios by Timeline

Year	Base Scenario	Stream 1	Stream 2	Stream 3	Stream 4
1	SWA 1	SWA 3	SWA 1	South America 2	Spratleys
	Singapore	Southern Africa	Horn of Africa	Cameroon	Chad
		East Timor		Singapore	
2	Central Asia	Thailand	Central Asia	SWA 3	South America 1
	Thailand	Sierra Leone	Liberia	Thailand	Horn of Africa
				Haiti	
3	Horn of Africa	Spratleys	Balkans	Taiwan	SWA 2
	SWA 2	Haiti	Rwanda	S. Africa	Singapore
		Chad			
4	Thailand	Balkans	Singapore	Spratleys	Taiwan
	India	Egypt	Cameroon	Egypt	Haiti
			India		
5	SWA 2	SWA 1	SWA 2	SWA 1	SWA 2
	North Africa	North Africa	Taiwan	Rwanda	East Timor
		Liberia	Sierra Leone	East Timor	
6	Egypt	Central Asia	Spratleys	Central Asia	SWA 1
	Taiwan	India	Chad	North Africa	Rwanda
		Cameroon	Thailand	Singapore	
7+	MCO 1	MCO 1	MCO 1	MCO 1	MCO 1
	MCO 2	MCO 2	MCO 2	MCO 2	MCO 2

NOTE: SWA = Southwest Asia; MCO = major combat operation.

focuses attention on providing resources to support deterrent deployments. It ensures their funding while also placing major combat operation requirements in the planning, programming, budgeting, and execution (PPBE) process.

Figure S.2
"Most Likely" or Baseline Scenario

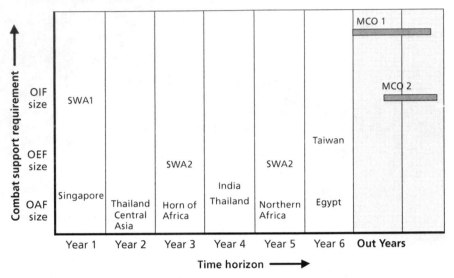

NOTE: OAF = Operation Allied Force; OEF = Operation Enduring Freedom;
OIF = Operation Iraqi Freedom.
RAND *MG421-S.2*

Selection of Existing Combat Support Bases

We solved the problem (i.e., we found the least-cost bases that would satisfy operational requirements) using existing forward support locations (e.g., Ramstein Air Base [AB]). The model selected 11 FSLs (see Table S.2). These locations represent the optimal locations to support the baseline scenario. Although the model was allowed to select from the four existing munitions preposition ships, none was chosen unless infrastructure expansion at the existing land-based FSLs was excluded from the solution. In that case, a single APF ship assigned to the Arabian Sea was used to compensate for the lack of storage space at the land-based FSLs.

We assessed the capabilities of the selected FSLs (see Table S.2) against the remaining four timelines. These FSLs, along with an additional site at Eielson Air Force Base (AFB), were able to meet the demand for three of the four additional streams, although with

Table S.2
Optimal Existing FSLs to Support the Baseline Scenario

Ramstein AB, Germany	Seeb, Oman
Sigonella AB and Camp Darby, Italy	Thumrait, Oman
RAF Mildenhall and Welford, UK	Kadena, Japan
Al Udeid AB, Qatar	Andersen AB, Guam[a]
Sheik Isa, Bahrain	Diego Garcia, UK
Masirah Island, Oman	

[a] The model did not select Andersen AB directly, mainly because of its remoteness and cost. However, the postoptimality analysis (Andersen is a "bomber island" with a large quantity of combat support resources) led to its selection.

increased transportation requirements and costs. However, for Stream 4, the 10-day initial operating capability (IOC) requirement had to be relaxed to 12 days for the South American deployment, and a single munitions ship (with Guam as its home base) appeared in the solution (see page 67).

Selection of Additional Combat Support Bases

The next step was to evaluate existing and potential FSLs against the baseline scenario and the four alternative streams of reality. We generated a list of potential FSL locations around the globe that could support a wide range of deployments; as before, the model selected an optimal list for the baseline scenario (the "most likely" scenario). The earlier 11 existing sites presented in Table S.2 remained in the solution (i.e., the model selected them again), along with five new sites in Europe and Asia: Incirlik, Turkey; Clark AB, Philippines; Paya Lebar, Singapore; U-Tapao, Thailand; and Balad, Iraq. It should be noted, however, that the list in Table S.2 is by no means sacrosanct, and alternative sites may provide the same capability at a similar or marginally greater cost. In particular, Souda Bay, Greece; Akrotiri, Cyprus; Constanta, Romania; or Burgas, Bulgaria, may be suitable alternatives to Incirlik, Turkey. In addition, some realignment of existing sites may be more efficient and effective than current sites. For example, the port of Salalla in Oman could be used to meet some requirements

met by Seeb or Thumreit with lower cost and less time than the current sites. The new combination of existing and potential FSLs offers
about 30 percent savings in total cost by reducing the overall transportation cost to the system (see page 69).

Figure S.3 illustrates the final results from the combination of
the baseline scenario and the four other streams of reality. This figure
also shows the locations of the other candidate sites that were not selected by the model. It and the accompanying Table S.3 divide these
locations into Tier 1 and Tier 2 categories. We use the label "Tier 2
FSLs" for a set of FSLs that require a more detailed consideration as
potential sites. They may also have appeared in the solution as a result
of one or two individual deployments, and therefore their role is
closely fixed to the nature of those particular deployments. Additionally, all the Tier 2 FSLs (with the exception of Puerto Rico) have uncertain political futures or limited internal capabilities. Iraq, for
example, falls in this category, but its location for support of many

Figure S.3
Supporting Global Deterrence Using a Global Set of Oveseas Bases

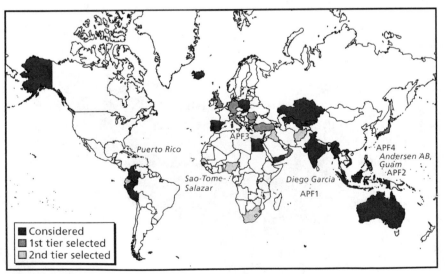

Table S.3
Global Set of Overseas Bases

Tier 1	Tier 2
Al Udeid AB, Qatar	Louis Botha, South Africa
Andersen AB, Guam	Bagram, Afghanistan
Diego Garcia	Baku, Azerbaijan
Kadena, Japan	Roosevelt Roads, Puerto Rico
Masirah Island, Oman	Tocumen, Panama
Mildenhall and Welford, UK	Cotipaxi, Ecuador
Ramstein, Germany	Sao Tome/Salazar, Sao Tome
Seeb, Oman	Kaduna, Nigeria[b]
Sheik Isa, Bahrain	Balad, Iraq
Sigonella and Camp Darby, Italy	
Thumrait, Oman	
Clark AB, Philippines	
Incirlik, Turkey	
Paya Lebar, Singapore	
U-Tapao, Thailand	
Souda Bay, Greece[a]	

[a] Alternatives to Souda Bay, Greece, are Akrotiri, Cyprus; Burgas, Bulgaria; or Constanta, Romania.
[b] An alternative to Kaduna, Nigeria, may be Dakar, Senegal.

operations makes it invaluable. However, we emphasize that the focus should not be on a particular latitude and longitude but rather on a particular region. Balad, Iraq, would be suitable if all the issues of security and long-term political amenities were resolved. If the uncertainties continue, then an alternative location in the region with similar capabilities should be considered (see page 75).

Figure S.4 presents the costs for the base scenario and all four streams. For each stream the expanded set of FSLs offers the same capability at a reduced overall cost to the Air Force. Note especially that the set of existing land-based FSLs could not support Stream 4 requirements and required that the IOC deadline be extended from 10 to 12 and also required the use of an APF munitions ship.

Figure S.4
Total Cost of Supporting All Scenarios Using Existing and Expanded Set of FSLs

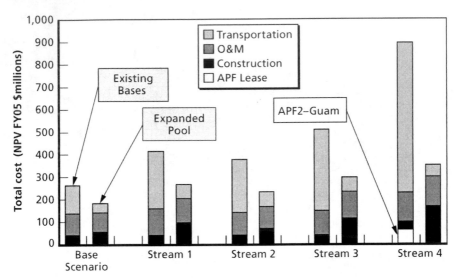

NOTE: Existing land-based FSLs cannot support Stream 4 requirements; the IOC deadline must be extended from 10 days to 12. NPV = net present value.
RAND MG421-S.4

However, when we selected from the expanded set of land-based FSLs, the need for the afloat option disappeared. The advantage of the global basing option is not limited to cost and encompasses a more efficient use of multimodal transportation. For each stream, the model was able to make better use of trucks and high-speed sealift for the expanded pool of bases, yielding about 50 percent less airlift usage without compromising operational requirements (see page 77).

Recommendations

We make the following recommendations based on our analysis of overseas combat support basing options:

   Using a global approach to select combat support basing locations is more effective and efficient than allocating resources on a

regional basis. One of the strengths of the analytic framework chosen is the lack of regional command boundaries. We are able to look at all regions of the world simultaneously with operations occurring in various locations at the same time, thereby extracting the most efficient solution without adversely compromising the capability needs of a particular region. Currently, the Air Force lacks a focal point for managing its investment in global infrastructure. Combatant Commanders influence their assigned warfighting units, which in turn influence Air Force investments on a regional basis, but there is no central organization that has the overall responsibility to investigate how these regional capabilities interact to provide global force projection capabilities. One option to overcome this shortfall would be the creation of a centralized Air Force planning and assessment group at the Air Staff. Because the potential scenarios impacting U.S. interests are constantly shifting, such a group needs to continually revise the model inputs and rerun these computer models, to ensure that the logistics posture is well suited to the current environment. This group might also have the budgeting and Programmed Objectives Memorandum preparation responsibilities associated with global logistics infrastructure (see page 83).

Political concerns need to be addressed in any decision about potential overseas basing locations. For instance, while an APF is much more expensive than alternative land-based storage options and may suffer from increased risk in deployment time, it may be necessary to consider the APF option because it offers more flexibility if access is denied. Additionally, countries like Iraq are continually selected by the model because cost and time are its major driving criteria. However, the uncertainty surrounding the future of Iraq (and similar countries) should force us to pause and consider alternative sites that may be less desirable mathematically but offer a higher probability of access and stability (see page 84).

Closer attention should be paid to Africa both as a source of instability and as a possible location for combat support bases. Africa, with its potential as a future source of oil combined with the uncertain future of many of its nation states, requires a great deal of attention from policymakers. Northern and sub-Saharan Africa continue to be plagued by civil wars, ethnic or clan-based conflicts, and/or severe economic disasters. There is a greater likelihood that terrorists may seek haven in the remote areas of Africa because of the continued U.S. military presence in the Middle East and Southwest Asia. Also, the geopolitical importance of the region, with its high levels of oil production, makes it an area of interest to the United States. If deployments to the region increased in the future, the current set of bases would not support those operations. Possible FSL locations in Africa could support operations across the entire southern half of the globe. Although the initial construction costs for these bases would be high, the costs would be quickly offset by the reductions in transportation costs. As an initial phase, we recommend closely evaluating western regions of Africa, with particular attention to Nigeria, Sao Tome/Salazar, South Africa, and Senegal. The development of African FSLs could be tied into other foreign policy and outreach initiatives in Africa, such as the NATO Mediterranean Dialogue country relationships with Algeria, Mauritania, Morocco, and Tunisia (see page 84).

Some Eastern European nations should be considered as serious candidates for future overseas bases. The potential for continued conflicts in central Asia and the Near East has made many of the countries in the eastern part of Europe very attractive as potential storage locations for WRM. The appeal of this region has been further heightened by the inclusion of some of these countries in the European Union (EU) and NATO, combined with the lower cost of living and the relatively high professional labor market. Romania and Bulgaria in Eastern Europe, along with Mediterranean locations such as Greece and Cyprus, form an appealing region that would allow easy access to both the U.S. Central Command (CENTCOM) and the U.S. European Command (EUCOM). These locations are especially attractive because they allow for multimodal transport options, using Black Sea ports for Romania and Bulgaria (assuming passage through the Bosporus Strait in Turkey to the Mediterranean). Poland and the Czech Republic, although very accommodating to U.S. efforts in the current operations, are located relatively far from the potential deployments that were considered in this report. Also, the

Czech Republic is a landlocked state, and while Poland has significant coastline on the Baltic Sea, these ports do not allow for rapid transport to the regions of U.S. Air Force (USAF) interest. In terms of transportation time and cost, neither Poland nor the Czech Republic offers savings versus the existing installations in Germany, and either would require a substantial investment in transportation infrastructure to attain the current capability levels in Germany.

Southeast Asia offers several robust options for allocation of combat support resources. The remoteness of Guam and Diego Garcia from most potential conflicts in the region requires the consideration of other locations in the Pacific. The geographical characteristics of the U.S. Pacific Command (PACOM) put a heavy reliance on airlift and possibly fast sealift. Most of the current U.S. bases are located in Japan and the Korean Peninsula with the main purpose of supporting the Korean deliberate plan. To support other possible contingencies, we propose a closer examination of three locations: Thailand, Singapore, and the Philippines. Each of these locations offers a host of options for the Air Force, including storage space, adequate runway facilities, proximity to ports, and strategic location. Darwin, Australia, has many of the desired attributes for an overseas combat support base, but its remoteness to any potential conflict makes it a comparatively poor choice.

Potential future operations in South America may be greatly constrained unless additional infrastructure in the region is obtained. In our analysis, a large South American scenario obtained from the Defense Planning Scenarios overstressed the system of existing facility locations, preventing the satisfaction of a 10-day IOC deadline, even with the use of APF ships. While the states of South America are relatively stable, the recent difficulties in Ecuador, Bolivia, and Venezuela demonstrate the potential volatility of the region. As with Africa, future U.S. intervention cannot be discounted owing to significant U.S. interests in the region's oil supply. Although the current combat support infrastructure is sufficient for small-scale operations such as drug interdiction, an expanded combat support presence would facilitate larger-scale operations in the region (see page 86).

A multimodal transportation option is the key to rapid logistics response. RAND has shown in several earlier reports (Amouzegar et al., 2004; Vick et al., 2002) that overreliance on airlift may in fact reduce response capability because of throughput constraints and lack of airlift. A comprehensive mobility plan should include a combination of air, land, and sealift. Judicious use of trucks and high-speed sealift in fact may offer a faster and less expensive way to meet the Air Force's mobility needs (see page 86).

Acknowledgments

Many people inside and outside the Air Force provided valuable assistance and support to our work. We thank Lieutenant General Donald Wetekam, AF/IL, who sponsored this research and continued to support it through all the phases of the project.

We are grateful to our project officers, Brigadier General Ronald Ladnier, AF/ILG, and Colonel Chris O'Hara, AF/ILGX. Their comments and insights have sharpened the work and its presentation.

At United States Air Forces in Europe (USAFE), we were fortunate to have great support from the USAFE/CC and his staff. In particular, we thank General Gregory Martin, AFMC/CC (formerly USAFE/CC), and Brigadier General Jay Lindell, USAFE/LG. At PACAF, Brigadier General Polly Peyer, PACAF/LG, and her staff provided tremendous assistance in our data collection. We thank Brigadier Generals Art Morrill, AF/ILP, and Dave Gillett, AF/ILM, for their support, feedback, and encouragement at the Air Staff and when they were serving at PACAF/LG and USAFE/LG, respectively. We also benefited from conversations with Major General Robert Elder, AWC/CC (formerly CENTAF/CV).

Many individuals supplied us with detailed information on specific programs. Among them, we especially thank Mr. Laine Krat, AF/ILGD, and Major John Bell, AFIT, for assisting us with our FSL data gathering; the current and former CENTAF/AF staff, including Colonel Michael Butler, Colonel Duane Jones, and Major Dennis Long for evaluating the early versions of our model; and Lieutenant Colonel David Frazee, the Executive Coordinating Officer for WRM in CENTAF (located in Muscat, Oman), who facilitated our visit to the storage sites in the Middle East.

We benefited greatly during our project from the comments and constructive criticism of many RAND colleagues, including (in alphabetical order) Edward Chan, Lionel Galway, Jim Masters, Patrick Mills, Kip Miller, and Don Snyder. We also benefited from careful reviews by David Oaks and Paul Dreyer. Finally, we thank Fran Teague for her assistance in the many revisions of this report.

As always, the analysis and conclusions are the responsibility of the authors.

Abbreviations

AB	air base
ACS	Agile Combat Support
ACN	aircraft classification number
AEF	Air and Space Expeditionary Force
AFB	air force base
AF/IC	Air Force Deputy Chief of Staff for Installations and Logistics
APF	afloat preposition fleet
APS	afloat preposition ship
BEAR	Basic Expeditionary Airfield Resources
CBP	capability-based planning
CENTCOM	U.S. Central Command
CIRF	centralized intermediate repair facility
COCOM	Combatant Commander
CONUS	Continental United States
CRAF	Civil Reserve Air Fleet
CSL	cooperative security location
DoD	Department of Defense
EU	European Union
EUCOM	U.S. European Command
FOC	full operational capability

FOL	forward operating location
FSL	forward support location
FSS	Fast Sealift Ship
FY	fiscal year
FYDP	Future Years Defense Program
GAMS	General Algebraic Modeling System
GUI	graphical user interface
HSS	high-speed sealift
INCAT	International Catamaran
IOC	initial operating capability
JFAST	Joint Flow and Analysis System for Transportation
LCN	load capacity number
MCO	major combat operation
MOG	maximum on ground
MSC	Military Sealift Command
MPMS	Multi-Period–Multi-Scenario
MTW	major theater war
NATO	North Atlantic Treaty Organization
NEW	net explosive weight
nmi	nautical mile(s)
O&M	operations and maintenance
OAF	Operation Allied Force
OEF	Operation Enduring Freedom
OIF	Operation Iraqi Freedom
PACOM	U.S. Pacific Command
PAX	passengers
PCN	pavement classification number
POM	Programmed Objectives Memorandum

PPBE	planning, programming, budgeting, and execution
PRV	property replacement value
ROBOT	RAND Overseas Basing Optimization Tool
RO/RO	roll-on/roll-off
RRDF	roll-on/roll-off discharge facility
SECDEF	Secretary Of Defense
SOF	Special Operations Forces
SWA	Southwest Asia
START	Strategic Tool for the Analysis of Required Transportation
TBM	theater ballistic missile
USAF	United States Air Force
USAFE	United States Air Forces in Europe
USTRANSCOM	United States Transportation Command
UTC	Unit Type Code
WRM	war reserve materiel

CHAPTER ONE

Introduction

Since the end of World War II, the United States has established and maintained a large number of overseas military bases, presently numbering more than 700 locations across the globe.[1] This massive presence has enabled the U.S. military to operate in every part of the world and respond to crises quickly. Although the genesis of permanent forward presence was established in the aftermath of World War II, the Korean War and the Cold War reinforced the importance of such bases. For more than four decades, these forward bases existed to serve one main goal: to prevent Soviet—and by extension North Korean—aggression against U.S. interests. The end of the Cold War ushered in a new era in the global security environment, causing a shift in the force posture toward a focus on supporting two major regional conflicts while simultaneously reducing the overall size of the military.

The end of the Cold War, however, did not reduce the burden on U.S. forces. In fact, in the last decade of the twentieth century the United States carried a significant portion of the security and peacekeeping responsibilities around the globe.[2] The U.S. Air Force

---

[1] One hundred fifty-six countries host U.S. troops, with U.S. bases in 63 countries. Since September 11, 2001, at least 13 new bases in 7 countries have been established. Overall, only 46 countries in the world do not have any U.S. military presence. For more information, see Eyal (2003). Also see U.S. Congressional Budget Study (2004) and DoD (2004).

[2] For example, in fiscal year 1999, U.S. Air Force operations included 38,000 sorties associated with Operation Allied Force, 19,000 sorties to enforce the no-fly zones in Iraq, and about 70,000 mobility missions to over 140 countries (see Sweetman, 2000). As of August

(USAF) has been called on to make numerous overseas deployments, many on short notice—using downsized Cold War legacy force and support structures—to meet a wide range of mission requirements associated with peacekeeping and humanitarian relief, while maintaining the capability to engage in major combat operations, such as those associated with operations over Iraq, Serbia, and Afghanistan. A recurring challenge facing the post–Cold War Air Force has been its increasing frequency of deployments to increasingly austere locations.[3]

Creation of the Air and Space Expeditionary Force

In response to the post–Cold War threat environment, the U.S. Air Force developed the Air and Space Expeditionary Force (AEF) concept, which has two primary goals.[4] The first goal is to improve the ability to deploy quickly from the Continental United States (CONUS) in response to a crisis, commence operations immediately on arrival, and sustain those operations as needed. The second is to reorganize to improve readiness, better balance deployment assignments among units, and reduce uncertainty associated with meeting deployment requirements. The underlying premise is that rapid deployment from CONUS and a seamless transition to sustainment can substitute for an ongoing U.S. presence in-theater.

To implement the AEF concept, the Air Force created ten Aerospace Expeditionary Forces, each comprising a mixture of fighters, bombers, and tankers.[5] These ten AEFs respond to contingencies on

---

2003, 16 of the Army's 33 combat brigades were operating in Iraq, and only about 7 percent of the approximately 160,000 coalition soldiers in Iraq were non-American.

[3] This point of Air Mobility Command deployments from 1992 to 2000 is discussed in Brunkow and Wilcoxson (2001).

[4] The Air Force defines "expeditionary" as conducting "global aerospace operations with forces based primarily in the U.S. that will deploy rapidly to begin operations on beddown" (USAF, *EAF Factsheet*, June 1999).

[5] Henceforth, when it is clear from the context, we will use AEF to represent both the concept and the force package.

a rotating basis: for 120 days, two of the ten AEFs are "on call" to respond to any crisis needing air power. The on-call period is followed by a 16-month period during which those two AEFs are not subject to short-notice deployments or rotations.[6] In the AEF system, individual wings and squadrons no longer deploy and fight as full and/or single self-sustained units as they did during the Cold War. Instead, each AEF customizes a force package for each contingency consisting of varying numbers of aircraft from different units. This fixed schedule of steady-state rotational deployments promises to increase flexibility by enabling the Air Force to respond immediately to any crisis with little or no effect on other deployments.

The dramatic increase in deployments from CONUS, combined with the reduction of Air Force resource levels that spawned the AEF concept, has equally increased the need for effective combat support.[7] Because combat support resources are heavy and constitute a large portion of the deployment tonnage (as shown in Figure 1.1), they have the potential to enable or constrain operational goals, particularly in today's environment, which depends greatly on rapid deployment.

Much of the existing support equipment is heavy and not easily transportable; deploying all of the support for almost any sized AEF from CONUS to an overseas location would be costly in terms of both time and airlift. As a result, the Air Force has focused attention on streamlining deploying unit combat support processes, reducing the size of deployment packages, and evaluating different technologies for making deploying units more agile and more quickly

---

[6] Beginning with AEF Cycle 5 (September 2004), the baseline deployment was changed from 90 days to 120 days, and the AEF cycle was changed from a 15-month rotational cycle to a 20-month cycle. For more information on AEF cycles see https://aefcenter.acc.af.mil/. There is, however, an expectation that for some stressed career fields, such as military police, there will be longer deployment periods of greater frequency (see Chief's Sight Picture: Adapting the AEF—Longer Deployment, More Forces, 6 July, 2004, http://www.af.mil/media/viewpoints/adapting_aef.html, as of September 22, 2004).

[7] Air Force Doctrine defines *combat support* to include "the actions taken to ready, sustain, and protect aerospace personnel, assets, and capabilities through all peacetime and wartime military operations" (USAF, 1997).

Figure 1.1
Substantial Support Footprint for Air and Space
Power (OIF Air Force Deployments)

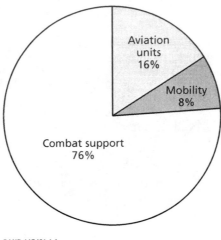

deployed and employed. Decisions on where to locate intermediate maintenance facilities, such as Jet Engine Intermediate Maintenance (JEIM) shops, and non-unit heavy resources (i.e., those not associated with flying units, such as munitions, shelters, and vehicles) are significant drivers of employment timelines.[8]

Events of September 11, 2001: A New Catalyst for Change

The events of September 11, 2001, and the subsequent conflicts in Afghanistan and Iraq propelled a second shift in the security environment in less than ten years. Although the Department of Defense (DoD) was reviewing its overseas basing options before that date, these events were a major catalyst for changes in military force pos-

---

[8] Since the end of the Cold War and the inception of the AEF concept, RAND has worked with the Air Force to determine options for intermediate maintenance—and for combat support as a whole—that could meet the Air Force's changing needs. For some of these works see, e.g., Tripp et al. (1999); Galway et al. (2000); Killingsworth et al. (2000); Peltz et al. (2000); Amouzegar, Galway, and Geller (2002); Feinberg et al. (2001); Amouzegar, Tripp, and Galway (2004); and Amouzegar et al. (2004). For a comprehensive review of RAND agile combat support work see Rainey, et al. (2003).

ture. DoD force planning focused on four major categories (DoD, 2001):

- defense of the U.S. homeland
- deterrence of aggression and coercion in critical regions of the world
- swift defeat of aggression in overlapping major conflicts
- conducting a limited number of smaller-scale contingency operations.

It has been clear for some time that U.S. defense policymakers can no longer plan for a particular deployment in a specific region of the world because the geopolitical divide of the last century has been replaced with a security environment that is more volatile.[9] One of the many lessons of the past decade has been the unpredictability of the nature and the location of the conflicts. In the conflict in Serbia, the USAF and coalition air forces played a major role in driving the Serbian forces from Kosovo. The common thought of the day was that all future conflicts would be air dominated. The events of September 11, 2001, and the consequent U.S. reprisal against Al-Qaeda in Afghanistan, Operation Enduring Freedom, reemphasized the importance of asymmetric warfare and the fundamental role of Special Forces. These events, however, have not lessened the need for a powerful and agile aerospace force as the USAF flew long-range bombers to provide close air support to the Special Operations Forces (SOF) working with the indigenous resistance ground force in Afghanistan, far from existing U.S. bases. In Operation Iraqi Freedom, the Air Force played a substantial role throughout the conflict—from its initial role of suppressing and disabling the Iraqi command and control and the air defense system to providing close air support in urban environments (Tripp et al., 2004; Lynch et al., 2005).

Although past conflicts and engagements may not be repeated in the same manner in the future, we can leverage our understanding of

---

[9] This is not to diminish threat-based analysis but rather to consider it as part of a complete solution.

those events to help shape our planning for the future. Moreover, we can focus on the characteristics of past events to create a broad set of alternative realities for the future environment.

New Combat Support Planning Strategy for the 21st Century: Deterrence in an Age of Persistent Global Insurgency and Counterinsurgency

In the current national security arena, the focus has shifted away from the post–Cold War paradigm of preparing for nonrecurring major regional conflicts. Instead, the focus is on ongoing and succeeding engagements and reconstitutions to deter aggression and coercion throughout the world, both by state and nonstate actors, while preparing to engage and succeed in major theater wars (MTWs).[10]

For more than 50 years, U.S. *deterrent strategy* was based on assured destruction, i.e., informing potential adversaries that the United States had overwhelming nuclear capabilities and could assure the destruction of state actors should they launch a first strike against it. The intent of this strategy was to ensure deterrence by making the thought of a first strike inconceivable. This nuclear deterrent strategy was accompanied by the creation of a large standing conventional force that could be employed to win conventional wars against the Soviet Union and North Korea (even if supported by the People's Republic of China). The strategy resulted in the development of large "standing capabilities" that could be augmented quickly by reserve components. Other contingencies were deemed to be a lower-intensity version of the MTW scenarios. The sole purpose was to develop an intimidating force with the expectation of avoiding an all-out engagement.

Today, the threat facing U.S. interests is different, and so are the necessary deterrent capabilities. As in the past, nuclear deterrence

---

[10] This shift can be mapped from the Cold War era of planning for a single war in Europe, to the post–Cold War two-MTW scenarios, to the present state of the world with multiple and sometimes shadowy adversaries.

continues to be vital against possible state actors, but a different conventional deterrent strategy is essential for the foreseeable future. In today's environment, rapid global force projection capability is needed to deter aggression and, if that fails, to take quick action to defeat state and nonstate actors. This deterrence concept involves the continuous and rapid projection of forces, primarily from CONUS, to sites in unstable regions around the world. This concept has the dual objectives of promoting stability and demonstrating that the United States can project power and destroy or diminish the capability of terrorist groups or state actors should they threaten U.S. or allied interests in the region. In short, a shift is needed from the paradigm of building capabilities to avoid a nuclear war to one of continuous use of forces to deter aggression and coercion.

The Effect on Programming and Budgeting

This deterrent framework changes the economic emphasis in the Programmed Objectives Memorandum (POM) process. POM is the critical tool of the planning phase in the Planning, Programming, Budgeting, and Execution (PPBE) System.[11] The PPBE process is the current system for creating the DoD's contribution (including that of all the services) to the presidential budget. The system divides the budget-building process into four phases:

- **Planning.** Assesses capabilities, reviews threats, and develops guidance.
- **Programming.** Translates planning guidance into achievable packages in a six-year future defense program.
- **Budgeting.** Tests for feasibility of programs and creates budgets.
- **Execution.** Develops performance metrics, assesses output against planned performance, and adjusts resources to achieve the desired goals.

---

[11] In practice, the programming and budgeting phases are combined and POM submissions are developed in conjunction with budget estimate submissions, the primary tool of the budgeting phase.

The Major Commands (MAJCOMs) submit these programs in the form of a POM to a body on the Air Staff called the Air Force Corporate Structure. The resources covered in the POM refer mainly to manpower, facilities, weapon systems, and operating funds.[12]

The new deterrence framework supports an expansion of the POM purview to include the resources necessary to support the routine deployment of forces to exercise sites. The expanded POM should also include resources to fight and win contingencies should deterrence fail, as the current POM does. This emphasis would cause more attention to be paid to deterrence exercises, along with the timing and resources necessary to support these exercises. The actual costs of engaging in contingencies should deterrence fail is not part of deterrence, and funding to engage in contingency activities would need to be handled on a case-by-case basis, as it is today.

The new security environment also places emphasis on a global view of combat support resources and their placement. This global view is the purview of the Secretary of Defense (SECDEF) and Headquarters Air Force. Although the Combatant Commanders (COCOMs), given their respective regional responsibilities, will continue to be interested in receiving support in their regions, their requirements need to be considered from the larger global vantage point. Therefore, the Air Staff (and the Joint Staff) should conduct a quantitative and objective analysis of the consequences of programming decisions for placement of limited resources. Furthermore, COCOMs and others would be interested in how political constraints—which either restrict some storage locations or force the use of other locations—are likely to impact effectiveness and costs.

Development of a Multi-Period–Multi-Scenario Combat Support Planning Methodology

The Air Force's new role will inevitably include a commitment to multiple, possibly overlapping, engagements in diverse geographical areas with varying degrees of operational intensity. Some of these en-

---

[12] For more information on the PPBE process, see Snyder et al. (2005).

gagements (e.g., drug interdictions) will occur multiple times over a short time horizon. To capture the nuances of the multifaceted continuous deterrent environment, we must integrate temporal and spatial elements with other parameters, such as combat support capability and costs. These parameters are captured in a new planning methodology in which several likely deployment scenarios, from small-scale humanitarian operations to major regional conflicts, are considered.

After the list of scenarios is generated, the *sequencing* and *recurrence* of these scenarios should be outlined. For any given scenario, decisions should be made regarding its likelihood of occurrence over time (e.g., a given scenario may be highly unlikely over the next five years, but considerably more feasible 20 years out), its interrelationship with other scenarios (e.g., Scenario A may likely occur simultaneously with Scenario B), and the likelihood that it will recur (e.g., a given scenario might repeat itself ten years out). We have coined the term *Multi-Period–Multi-Scenario* (MPMS) to describe this planning methodology. This methodology is a major departure from the current war planning mindset. Previously, whether planning for nuclear warfare against the Soviet Union or for large-scale conventional war in the Near East, U.S. analysts were planning for one large conflict that would occur only once and that would change the defense environment so greatly that plans for out-years following this conflict would no longer be valid.

A Need for New Combat Support Basing Options

The current overseas basing postures that are concentrated in Western Europe and Northeast Asia may be inadequate for the 21st century because potential threats have transcended the geopolitical divide of the Cold War era. The events in Southwest Asia prior to OIF, the difficulties of securing basing access in Turkey during OIF, and the denial of overflight rights from countries that opposed the war in Iraq, such as Austria, have further emphasized the importance of alternative forward operating and support locations.

In the European theater, there has been an interest among recent and aspirant NATO and EU member countries in being potential hosts for U.S. military combat and support forces. The Supreme Allied Commander Europe (SACEUR), General James Jones, U.S. Marine Corps, has been interested in reevaluating bases in Europe for some time, and on February 26, 2003, the House Armed Services Committee heard testimonies on "U.S. Forward Deployed Strategy in the European Theater." At the same meeting, a representative of the American Enterprise Institute argued that some of the existing force bases in Germany should be moved to Poland, Romania, and Bulgaria (House Armed Services Committee, 2003).

As mentioned earlier, the idea of a new basing strategy has been circulated for some time both inside and outside of the Pentagon.[13] Table 1.1 shows a summary of U.S. overseas bases for all the services. Although the United States has bases in many countries, the vast majority of the large installations are concentrated in support of the two-MTW concept (i.e., one in Western Europe and one in Northeast Asia). Current European bases—home to thousands of U.S. troops and their families—may be far from potential conflicts. Furthermore, because of economic differences across countries, costs may be higher in Western Europe than in Eastern and Central European countries. Public support for the U.S. presence may be eroding in Germany, and, to a lesser extent, in the United Kingdom.[14]

Arguably the most dramatic effect of OIF will be felt in the Near East, where a major threat in the region has been eliminated and an opportunity may have opened for transforming Iraq into the United States' chief regional ally. In 2003, the Pentagon announced a plan for withdrawing its forces from Saudi Arabia, ending a 20-year military presence; it is now looking for alternatives in the region in

---

[13] We have had several internal discussions and have informally spoken about examining new forward support locations in Central and Eastern Europe in briefings to senior Air Force leaders in recent years.

[14] This was particularly true during OIF, an unpopular conflict in Western Europe.

Table 1.1
Summary of Overseas Bases, FY 2004 Baseline Data

Service	No. of Large Installations	No. of Medium Installations	No. of Small Installations	Total
Army	1	8	364	373
Navy	5	4	188	197
Air Force	7	7	255	269
Marine Corps	2	2	19	23

SOURCE: DoD (2004).
NOTE: Large Installation = Total property replacement value (PRV) of greater than or equal to $1.5 billion. Medium Installation = Total PRV between $829 million and $1.5 billion. Small Installation = Total PRV less than $828 million.

which to base some of its troops and support equipment (Burger, 2003). The U.S. Air Force has removed 50 warplanes from Incirlik, Turkey—ending Operation Northern Watch, the decade-long enforcement of the no-flight zone over northern Iraq.

This new strategic realignment also affects Asia, where the Pentagon is considering or has already decided to shift some troops from their long-standing major bases in Japan and Korea and to establish smaller bases in such countries as Australia, Singapore, or Malaysia ("U.S. to Realign Troops in Asia," 2003).[15]

This transformation is partly a continuation of a readjustment that began over a decade ago at the end of the Cold War, partly due to the events of September 11, 2001, and their aftermath and partly because of the new realignment in world security as the result of the operation in Iraq. But whatever the reason or cause, it has produced a spectacular change of strategy that requires thoughtful planning and analysis. The old alliances—the North Atlantic Treaty Organization (NATO), the Australia-New Zealand-U.S. (ANZUS) partnership, and the Southeast Asia Treaty Organization (SEATO)—may survive for years to come, but their roles and importance could diminish as

---

[15] On June 5, 2003, the Pentagon announced withdrawal of U.S. troops from the Demilitarized Zone (DMZ) in Korea.

the member nations assign differing, probably competing, values to the merits of various ventures that the United States considers important. Given the fluid nature of the threat and potentially differing levels of support from allies, the USAF may wish to consider a number of new options in supporting its forces.

One strategy has been to establish cooperative security locations (CSLs) for combat support.[16] As an example of such a location, 200 airmen from throughout Europe set up a temporary KC-135 tanker base in Bulgaria during OIF. The Bulgarian military and local police provided most of the security, local contractors provided fuel and meals, and the Air Force security forces guarded the planes (Simon, 2003). This scheme has the advantage of precluding any political ramifications of permanently stationing American troops on foreign soil, as has been evidenced in Saudi Arabia and South Korea. Other new facilities would be considerably smaller and more austere than current military bases, such as the one in Ramstein, Germany. As General Jones said on April 28, 2003, in Washington, D.C., these will be locations "that you can go to in a highly expeditionary way, land a battalion, train for a couple of months with a host nation, if you will, or part of an operation, and then leave and then come back maybe six months later" (Burger, 2003).

Organization of This Report

In Chapter Two, we discuss the geopolitical environment in which the U.S. military has to operate. We start with a brief discussion of U.S. military operations since 1990 and then present a review of the state of affairs in the Near East, the Asia-Pacific region, Central Asia, South America and the Caribbean, Europe, and Africa.

In Chapter Three, we develop a set of deployment principles that would facilitate the evaluation of various options for combat support basing. In this chapter, we highlight the different deployment

---

[16] *Cooperative security locations* is the U.S. Air Force's new terminology for such bases.

characteristics and show how they are incorporated into our "streams of reality."

Chapter Four deals with the combat support factors essential to the selection of alternative forward support locations. These factors include base accessibility, vulnerability, and capability. This chapter also presents alternative options to air transportation, including the use of high-speed sealift (HSS); discusses the capabilities of U.S. and other basing options; and concludes with a list of 50 potential sites for FSL location.

Chapter Five presents a detailed discussion of our analysis methodology including the development of a large-scale optimization model.

In Chapter Six, we illustrate the use of our analytic framework by presenting an analysis of global basing decisions including the evaluation of existing and potential land- and sea-based forward support locations. That chapter presents the results of several computational runs and shows different basing options that are robust across various deployments.

Chapter Seven presents our conclusions and recommendations for the combat support basing options.

CHAPTER TWO

The Geopolitical Environment

One of the United States' major defense policy goals is to deter threats and coercion against U.S. interests anywhere in the world. This multifaceted approach requires forces and capabilities that discourage aggression or any form of coercion by placing emphasis on peacetime forward deterrence in critical areas of the world. In addition, U.S. forces must maintain the capability to support multiple conflicts if deterrence fails (DoD, 2003b). U.S. interests are not only global but dynamic as well, particularly when the nation is confronted with emerging anti-access and area denial threats. The Air Force core competencies, such as agile combat support, global attack, and rapid global mobility, reflect these changes in the global threat environment. *Agile combat support* is defined as "the capability to create, protect, and sustain Air and Space forces across the full spectrum of military operations" (USAF, 2005). *Global attack capability* is defined as "the ability to engage adversary targets anywhere, [and] anytime." *Rapid global mobility* is defined as "the ability to rapidly position forces anywhere in the world" (U.S. Air Force, 2000).

The Air Force can rapidly airlift forces anywhere in the world if those forces are sufficiently small and if the airlift is not consumed by other requirements elsewhere. However, the United States' strategic policy goals and the reality of today's security environment require a capability that can project a continuum of power both swiftly and globally. Doing so requires a combat support system that has both agility and adaptability to support a broad range of potential engagements anywhere in the world. To help forecast potential combat support requirement for U.S. forces, we examine historical patterns of U.S. operations, current crises and conflicts, and potential engagements with terrorist groups and their state sponsors.

U.S. Operations and Exercises Since 1990

It has been more than a decade since the end of the Cold War, and in that period U.S. forces have been involved in numerous operations and conflicts.

Figure 2.1 illustrates some of the deployments since early 1990. Naturally, the number of operations does not reflect the intensity or the level of interest because certain conflicts have put more burdens on the armed forces than others. Although the United States does not respond to every crisis, the regions of the world in which these operations have been conducted reflect the strategic interests of the United States and its allies.

Figure 2.1
Sample of U.S. Operations and Exercises Since 1990

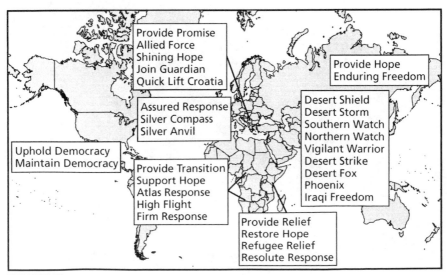

RAND MG421-2.1

Many of the deployments listed in Figure 2.1 occurred in regions where the United States has either a permanent support infrastructure (e.g., Europe) or a long-standing presence (e.g., the Near East). However, this figure indicates that a large number of recent deployments have required U.S. forces to enter new locations that had neither existing U.S. infrastructure nor a historical U.S. presence. Factoring in the relative paucity (by Western standards) of these locations' organic logistics infrastructure, these operations and exercises have frequently required deployments to bare bases, with the associated heavy use of combat support assets. Presently, the United States has avoided major conflicts south of sub-Saharan Africa or South America, even though it is interested in the future status of these regions.

Additionally, after September 11, 2001, the United States declared a war against terrorism and states that might support terrorist groups. This declaration was followed by an armed conflict in Afghanistan against the Taliban government and the Al-Qaeda group that operated in that country. These military operations, mainly made up of training, logistics support, and special operations, did not overly burden the combat support assets but were spread out across diverse regions of the world from Southeast Asia to Central Asia and beyond.

In the remainder of this chapter, we outline several potential military and nonmilitary operations in the Near East, the Asia-Pacific, Central Asia, South America, Europe, and northern and sub-Saharan Africa. The type and the location of potential operations were selected to reflect both historic U.S. involvement and potential locations where future conflicts might intersect with U.S. interests. We were also mindful of selecting a set of operations that would place varying stresses on the combat support system, so that we could evaluate a wide range of demands on the combat support requirements.

Near East

Despite the demise of the Baathist regime in Iraq, the U.S. military will continue to be involved in Iraq for the foreseeable future. Moreover, most nations in that region have authoritarian governments.[1] There is a potential for instability in many of these governments, which may not be able to cope with growth in popular unrest. Moreover, the potential growth of Islamic fundamentalism in many Arab countries may contribute to further volatility in this region. Although Iran may eventually become friendlier with the United States, its current system of government, with a powerful nonelected head of state, has severely hampered any movement toward normalization of relationships. Crises such as a sudden shock in Saudi Arabia would further change the security environment in the Persian Gulf and may consequently increase the importance of Iran's role in the region.[2] A destabilized Saudi Arabia and a potentially prolonged interruption of the flow of oil would have severe consequences for the United States and the global economy (Sokolsky, Johnson, and Larrabee, 2000). The most immediate threat may be the proliferation of weapons of mass destruction and the increase in insurgency movements.

In our analysis we continue to use different types of Southwest Asia (SWA) scenarios to simulate different-sized regional conflicts and to measure the combat support capabilities of alternative forward support locations. We also give attention to the extreme east of Africa, with the Horn of Africa playing an important strategic role in the status of the region as a whole.

---

[1] The main exception is Israel, which is democratic. Other countries, such as Iran, hold regular and relatively free elections, although Iran is influenced by its revolutionary ideals. Moreover, it is certainly too early to assess the outcome of Iraq's election of January 2005.

[2] For more information on Iran and its security strategies see Chubin (2002) and Byman et al. (2001).

Asia-Pacific

Over the past 50 years in the Asia-Pacific region, the United States has focused on the security of South Korea and has established support plans for containing North Korea. Although the United States will not be challenged by a "near-peer" for the foreseeable future, the potential for regional powers to develop capabilities to threaten U.S. interests exists. Asia may be the region where there could be large-scale challenges to the U.S. military (DoD, 2001).

China, in particular, may emerge as a more powerful naval force in the near future, challenging the U.S. Navy and Air Force dominance in the Pacific. Although China may not match the advanced military power of the United States, it could play an "asymmetric game" in the region by taking advantage of its vast coastline, as well as the long stretches of its rear base that reaches all the way to Central Asia.[3] China, in essence, may occupy the same role in the Pacific in this century that the Soviet Union played in Europe in the latter half of the twentieth century. Therefore, near-peer scenarios such as Taiwan-China or China-Russia (e.g., a Global Engagement IV–type model) can be used to assess the effect of potential FSLs in these very stressing scenarios. The sea-lanes of the South China Sea and the waters surrounding Indonesia are transited by nearly half of the world's merchant marine capacity. These areas are critical to the movement of U.S. forces from the Pacific to the Indian Ocean and beyond. Although the end of the Cold War has reduced the clear and immediate global military threat, the potential for both conventional and nonconventional threats still exists. One of the growing concerns is the threat of piracy and its connection to terrorism. Another issue is the overlapping claims by China and Taiwan over the South China Sea, with many Southeast Asian countries laying claims to the Spratley Islands (Khalilzad and Lesser, 1999). The distances in this region are

---

[3] As a simple illustration of the size of the U.S. and Chinese navies, consider the following: The U.S. Navy's warships have a total "full-load displacement" of about 2.9 million tons, whereas China's have less than 300,000 tons. The United States deploys 24 aircraft carriers (out of world's 34); China deploys none.

vast, and the density of U.S. basing and en-route support infrastructure is not as rich as in other important regions.

Other scenarios in this region would include antiterrorism activities in Indonesia and the Philippines as well as anti-insurgency operations in the Philippines.[4]

Central Asia

The support of some Central Asian countries in Operation Enduring Freedom, the ongoing U.S. military presence in Afghanistan, and the rich oil reserves of the Caspian Sea region—combined with potential conflicts in the Caucasus and Central Asia—have brought this region to the attention of many policymakers. However, the poor infrastructure of the Central Asian region provides a test to any combat support capability. Moreover, the trepidation of some NATO allies to project force into the region because of its proximity to Russia may put most of the burden on the United States (Sokolsky, Johnson, and Larrabee, 2000).

Some Central Asian countries may be able to play a role in supporting the USAF's continued efforts in Afghanistan or potentially in an Indian-Pakistani conflict. Furthermore, we are interested in measuring the effectiveness of the USAF's global storage and maintenance system in supporting a potential conflict in this region. For example, the tension between the ethnic Kazakh and Russian populations of Kazakhstan could trigger a civil war that may lead to the secession of the northern provinces of Kazakhstan or even Russian occupation of part or all of the country (Sokolsky, Johnson, and Larrabee, 2000).

---

[4] On August 5, 2003, a bomb exploded in Jakarta, and according to *The New York Times*, the Indonesian police and the Australian foreign ministry warned of the possibility of further attacks. Unfortunately, this prediction was realized when Bali was the site of another bombing on October 1, 2005.

South America and the Caribbean

The United States' continued efforts in antidrug activities in South America are likely to be the main focus for the military in this region.[5] Nevertheless, economic and political upheavals—such as the ones in Argentina and Venezuela, respectively—may require differing military roles for U.S. forces in the future. In this region of the world, planning concentrates on small-scale operations that would mostly involve Special Forces.

Europe

The United States has strong historical ties with Europe, with dozens of U.S. bases located across the continent. In the near term, it is hard to imagine any major conflicts in Europe such as the ones in the former Yugoslavian states that culminated with Operation Allied Force (OAF). Nevertheless, we will continue to include a variation of a Balkan scenario to test United States Air Forces in Europe (USAFE) combat support capabilities. In addition, we include a continued European role as a support command, as in OEF and OIF.

Africa

Northern and sub-Saharan Africa continue to be plagued with civil wars, ethnic or clan-based conflict, and/or severe economic disasters. The 2003 civil war in Liberia led to the deployment of Nigerian peacekeepers with a small U.S. force in the country.[6] In 2002, with the help of Britain and a large United Nations peacekeeping mission, the West African state of Sierra Leone emerged from a decade of civil

---

[5] At the urging of Peru and Colombia, President Bush may authorize the resumption of anti-drug surveillance flights over Colombia (*The New York Times*, August 6, 2003).

[6] The role of American troops was confined to assisting with logistics, reflecting the general uneasiness of the Pentagon over making a long-term commitment while U.S. troops are spread across the globe (*The New York Times*, August 5, 2003).

war. More than 17,000 foreign troops disarmed tens of thousands of rebels and militia fighters.[7] The Gulf of Guinea in West Africa has become a strategic interest to the United States because of its increased oil production.[8]

Recent developments in Northern Africa have been encouraging, with Libya pledging to abandon its pursuit of nuclear weapons. However, the continued threat of insurgencies in Algeria and Western Sahara may require future U.S. involvement in northern Africa. The countries of this region continue to be sources of Islamic fundamentalist groups, providing pools of recruits and staging areas for terrorist acts, most notably the Casablanca bombing of May 2003 and possibly the subway attack in Spain on March 11, 2004.

Across Africa, political instability and high levels of violence may continue to persist. The potential for extraction of large volumes of oil from African nations may add to their geopolitical importance. In this region of the world, we will concentrate our scenarios on humanitarian support requiring a small-scale aerospace force presentation.

---

[7] A short description of this action may be found on the BBC News Web site, http://news.bbc.co.uk/1/hi/world/africa/country_profiles/1061561.stm (last accessed October 7, 2005).

[8] Currently, 15 percent of U.S. oil is supplied by the Gulf of Guinea, a figure projected to grow to 25 percent by 2025 (*Baltimore Sun*, February 19, 2004).

CHAPTER THREE

Deployment Scenarios

The Department of Defense has made capability-based planning (CBP) a core tenet of its policy goals. CBP is based on the notion that the United States can no longer know, with a high degree of accuracy, what nation, combination of nations, or nonstate actors will pose a threat to vital U.S. interests (DoD, 2001). In the past planning environment, where the locations and capabilities of potential adversaries were known, the focus was on building an *optimal* combat support network to support these known threats. An unfortunate characteristic of optimally designed networks is that they often perform very poorly if the set of demands (locations and quantities) differs from the plan. The new planning environment, with its broad (and unclear) set of potential adversaries, calls for *robust* and *efficient* combat support networks that, while not necessarily optimal for any one scenario, perform well for a wide range of scenarios.

We made an assessment of the capability needs associated with the new planning environment by surveying a range of scenarios and generating a list of capability requirements. These scenarios consider such differing basing characteristics as basing availability, assurance of basing, and base security. They also include strategic factors, such as deployment distance, likely amount of strategic warning, deployment duration, and current Air Force reconstitution requirements. The scenarios have varying degrees of infrastructure richness, such as availability of fuel, communications, and transportation. An effective combat support system should be responsive to various types of demands or stresses. Indeed, the unpredictability of the future security environment requires the evaluation of support concepts across a broad range of combat and noncombat scenarios with varying degrees of intensity.

The Planning Process

Planning in such an environment calls certain geographically based assumptions into question. Previous RAND research (Amouzegar et al., 2004) has suggested that a global perspective be adopted in combat support basing planning. Forward positioning of combat support assets is recognized as key to the success of AEF rapid deployment. However, when potential combat operations occur near the boundary of several geographic commands (e.g., U.S. Central Command [CENTCOM] and the U.S. European Command [EUCOM]) or when operations exceed the capabilities of a single command, global resource allocation provides a strategy that allows for increased capability with lower overall costs through resource sharing. Another consideration supporting this global view is the realization that multiple storage locations are needed if access to desirable bases is denied, as was illustrated by the lack of access to Turkish bases in OIF.

To perform this analysis, it is necessary to create a list of scenarios that stress the combat support network. We considered a broad range of potential future engagements to identify a robust set of facility locations. Table 3.1 lists the regions we investigated, classified into three categories: major conflicts, exercises and other deterrent missions, and humanitarian and major operations other than warfare (MOOTW).[1] The table shows several major regional deployments that the SECDEF may believe are appropriate for developing the baseline case that will be used to develop the Program Objectives Memorandum and outlines contingency operations to measure the

---

[1] In developing the characteristics of deployments in these regions, we have relied on lessons learned from recent military activities while keeping in mind that past conflicts are merely indicators and not predictors of future requirements. Other sources of information are AF/XOX scenarios, Defense Planning Scenarios, and Khalilzad and Lesser (1999).

Table 3.1
Historical and Potential Regions of Conflict

Major Conflicts	OIF, OEF, OAF Asian littoral Korea
Exercises and Other Deterrent Missions	Southeast Asia South America Near East Central Asia Indian subcontinent Horn of Africa
Humanitarian and MOOTW	Southern Africa Central America and the Caribbean South Pacific Central and North Africa

effect on U.S. capabilities should deterrence fail. Careful attention needs to be given to the exercise and other deterrent missions category. This includes specifying the force packages and associated support forces that will need to deploy to deter aggression in the most likely deployments.

The goal in creating this list was to develop a set of scenarios in different regions of interest that stressed the system across a range of demands on the combat support network, with respect to both location and quantity of demand. These scenarios include potential military and nonmilitary operations in the Near East, the Asia-Pacific region, Central Asia, South America, Europe, and northern and sub-Saharan Africa. The motivation for considering these geographic re gions is discussed in Chapter Two.

We computed Air Force beddown using existing plans, historical data, and expert judgment. Given an aircraft beddown for each scenario, we estimated the combat support requirements needed at each forward operating location (FOL) using a RAND model, the Strategic Tool for the Analysis of Required Transportation (START) (Snyder and Mills, 2004). Although combat support is comprised of many consumable and repairable items, the focus of this report is on Basic Expeditionary Airfield Resources (BEAR), munitions, and rolling stocks (e.g., trucks) because they comprise the bulk of the items in the war reserve materiel (WRM) package.[2] Other commodities, such as fuel handling equipment, are necessary for successful FOL operations; however, these other commodities make up a small part of the WRM commodities that must be moved from storage sites to the FOLs. Prepositioned munitions are a special case because they are heavy and require special handling.[3] Because WRM is designed for use in austere locations, we were only concerned with deployments to Categories 3 and 4 FOLs for nonmunitions WRM and Categories 2, 3 and 4 for munitions using the following classification scheme (Galway et al., 2000):

- **Category 1:** Main Operating Base; fuel, infrastructure, munitions are available; full operational capability (FOC) within 24 hours of arrival[4]
- **Category 2:** Standby Base; some facility/support plus-ups required; fuel available; FOC within 48 hours of arrival
- **Category 3:** Limited Base; minimal infrastructure available; access to fuel; FOC within 48–96 hours of arrival
- **Category 4:** Bare Base; runway/taxiway/aircraft parking available; FOC within 72–120 hours of arrival.

Scenario Construction

To select a set of robust overseas combat support locations, we developed the following tenets for the construction of deployment scenarios:

---

[2] BEAR provides the required airfield operational capability (such as housekeeping or industrial operations) to open an austere or semi-austere airbase.

[3] For the remainder of this report, we will use the term WRM to mean BEAR, munitions, and rolling stock. When appropriate, we separate the commodities into munitions and non-munitions WRM.

[4] Time estimates for full operational capability are general and are dependent on closure of base operations capability packages.

- Although it is impossible to select combat support bases without specific operational deployments, the selection process should not be slaved to a particular deployment. For that reason, we do not seek to optimize the system for a handful of deployments alone.
- Combat support requirements should be dynamic and deployment scenarios should cast a wide geographical net in order to stress the combat support and transportation requirements.
- Deployments should be sequenced in time and space in order to evaluate physical reach and test the long-term effect of location and allocation of assets.
- To hedge against the uncertainty of the future security environment, multiple series of possible scenarios should be developed to test the robustness of the overseas combat support bases.

In the remainder of this section, we will present the details of the deployments developed for assessment of overseas bases. Figure 3.1 illustrates the variability of combat support requirements across the set of regions presented in Table 3.1 (excluding the major conflicts), in terms of the relative WRM combat support requirements. The vertical bars indicate the range of requirements for various potential deployments. On the Y-axis, the relative scale of recent deployments, in terms of WRM requirements, is noted to allow for comparison.

Note that in certain instances deterrence operations (e.g., South America) require greater combat support than a traditional major regional conflict such as Operation Allied Force. This is due to the fact that the FOLs considered in OAF are primarily well-developed locations with significant existing infrastructure, while the South American contingency requires deployment to more austere locations. Thus, although the OAF deployment may be much larger in terms of forces deployed, its requirements on WRM and munitions are less than those of the "smaller" deployment.

Korea is included in the analysis, but the resources allocated to Korea have been excluded from the model. The resources needed to

Figure 3.1
Relative Size of Combat Support Requirements Across Regions of Interest

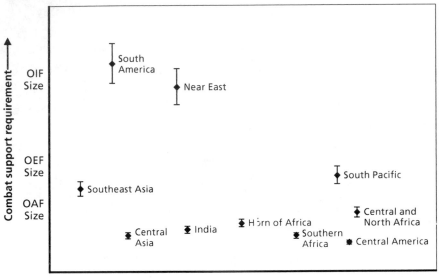

support exercises and operations in Korea have been quarantined from reallocation.[5]

In each region there may be several deployments, exercises, or deterrent missions, each with its own unique logistical characteristics. Table 3.2 lists some of the deployment missions we considered, along with their combat support requirements and number of forward operating locations. Note that the major combat operations (MCOs) are labeled as MCO1 and MCO2. These represent full-scale conventional warfare requiring large numbers of fighters and bombers; as such, they do not belong in the category of exercises and other deterrent missions. As we pointed out in the POM discussion of Chapter

---

[5] Although we have not done so in this analysis, it is possible to allow for reallocation of Korean resources if the threat from North Korea is deemed to have diminished.

Table 3.2
Deployment Characteristics

Contingency	No. of FOLs	BEAR (Short Tons)	Vehicles (Short Tons)	Munitions (Short Tons)
MCO 1	8	15,520	6,954	11,147
MCO 2	14	21,727	11,896	43,742
Balkans	4	2,612	1,567	1,326
Taiwan	4	5,194	2,189	12,778
SWA 1	4	16,200	6,104	19,470
SWA 2	1	5,843	2,348	10,044
SWA 3	4	7,098	3,307	3,509
South America 1	4	14,069	5,634	28,077
Central Asia	3	1,807	1,016	283
Spratley Islands	2	4,937	1,943	12,404
Thailand	2	4,411	1,057	1,855
Singapore	1	4,031	1,539	6,394
Eqypt	2	4,411	1,057	1,855
India	2	4,411	1,057	1,855
Southern Africa	2	1,807	1,119	0
Cameroon	2	1,807	1,119	0
Liberia	2	1,807	1,119	0
Sierra Leone	2	1,807	1,119	0
Haiti	1	1,681	1,119	0
Chad	2	1,807	1,119	0
Rwanda	1	1,681	1,119	0
East Timor	2	1,807	1,119	0
Northern Africa	3	3,811	2,050	0
Horn of Africa	2	3,686	2,050	0
South America 2	2	4,010	2,639	0

One, the actual cost of engaging in MCOs is not programmed because they are sourced through a different funding mechanism (e.g., supplemental). Nevertheless, it is essential to include MCOs in the analysis, since the combat support storage capacity and capability as well as the allocation of assets across the various locations will have a direct impact on the success of operations in a major conflict. The overseas bases must have enough capacity and throughput to support not just the exercises and the deterrent missions but any large-scale conflict as well. With the list of deployments defined, it is next necessary to outline the *sequencing* and *recurrence* of those deployments. We chose to schedule the deployments and contingencies into a scenario comprising a six-year time frame to align with the PPBE's six-year process.

Figure 3.2 shows a notional set of deterrence, exercises, and MCOs. Any combat support storage location that is selected must be able to support those deployments, including the possible MCOs. The deployments vary in terms of combat support requirements, as shown on the y-axis. Such a set of deployments, considered in unison, provides an integrated view of deterrence. This integrated, simultaneous assessment of deterrence and preparedness would be valuable to Combatant Commanders as they ensure that adequate exercises are conducted in *their respective areas of responsibility* to provide

Figure 3.2
Variations in Combat Support Requirements Across Time

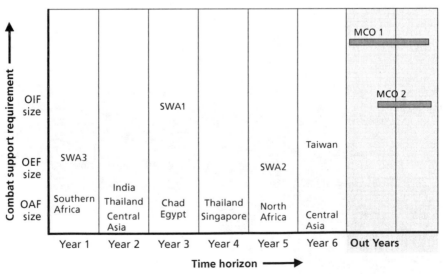

deterrence. The SECDEF and Joint Staff then may determine the baseline scenarios and include them in the Joint Strategic Capabilities Plan (JSCP) after considering inputs from the services (in this particular case, the Air Force) and the COCOMs.

To hedge against uncertainty, it is necessary to consider sets of potential scenarios (which we call *streams of reality*, or *timelines*) in order to identify a robust FSL posture. Given one timeline, we can use our optimization framework to identify an "optimal" FSL posture with respect to differing objectives, such as minimum cost, minimum deployment time, or minimum number of FSLs. Unfortunately, the truly optimal solution can only be computed if the future deployment schedule is known a priori. Therefore, we consider multiple scenarios and identify the *optimal* FSL postures for each stream individually. We then perform a *portfolio analysis* to identify FSL postures that perform well across every timeline and FSLs that provide robust solutions. The relative performance of this robust set of FSLs can be measured by comparing its performance versus the optimal solution for a given scenario. Ideally, multiple robust solutions should be identified to allow for other nonquantitative considerations (e.g., political constraints). This procedure is at the core of the MPMS.

The scenarios were scheduled into five streams or timelines according to the following set of rules. Each timeline was designed to include two major conflicts in order to sufficiently size the facilities to support regional conflicts specified in the planning guidance. However, as mentioned earlier, since the operational cost for wartime execution is not included in the POM, it can be assumed that these conflicts occur at the end of the six-year Future Years Defense Program (FYDP). The baseline scenario represents the most likely timeline and is used as the starting point for analysis. Each subsequent stream widens the geographical net and adds more stress to the combat support system. Table 3.3 on the following page contains the specific sequencing of deployments for the base scenario and each of the four timelines considered.

Table 3.3
Sequencing of Scenarios by Timeline

Year	Base Scenario	Stream 1	Stream 2	Stream 3	Stream 4
1	SWA 1	SWA 3	SWA 1	South America 2	Spratleys
	Singapore	Southern Africa	Horn of Africa	Cameroon	Chad
		East Timor		Singapore	
2	Central Asia	Thailand	Central Asia	SWA 3	South America 1
	Thailand	Sierra Leone	Liberia	Thailand	Horn of Africa
				Haiti	
3	Horn of Africa	Spratleys	Balkans	Taiwan	SWA 2
	SWA 2	Haiti	Rwanda	S. Africa	Singapore
		Chad			
4	Thailand	Balkans	Singapore	Spratleys	Taiwan
	India	Egypt	Cameroon	Egypt	Haiti
			India		
5	SWA 2	SWA 1	SWA 2	SWA 1	SWA 2
	North Africa	North Africa	Taiwan	Rwanda	East Timor
		Liberia	Sierra Leone	East Timor	
6	Egypt	Central Asia	Spratleys	Central Asia	SWA 1
	Taiwan	India	Chad	North Africa	Rwanda
		Cameroon	Thailand	Singapore	
7+	MCO 1	MCO 1	MCO 1	MCO 1	MCO 1
	MCO 2	MCO 2	MCO 2	MCO 2	MCO 2

NOTE: SWA = Southwest Asia; MCO = major combat operation.

CHAPTER FOUR

Identification of Candidate Locations

The previous chapter presented the notion of the MPMS concept, with its emphasis on planning for a series of deterrence scenarios. This chapter discusses the major constraining and contributing factors that influence the selection of overseas storage locations. The goal is to identify a robust set of overseas bases that can support a range of contingencies and deployments across the globe. Each deployment presents a different set of combat support factors and requirements:

- Strategic factors, such as warning time, affect the amount of equipment that can be deployed before an operation begins.
- The reconstitution condition of the Air Force impacts the amount of airlift available.
- The deployment distance between forward operating locations and potential bases (either CONUS or outside of CONUS [OCONUS]) affects the amount of airlift required (both tactical and strategic) and transportation time needed.
- The likely duration of the conflict affects the amount of equipment required within the theater.
- The infrastructure richness (including availability of fuel, communications, and commercial transport) at the FOLs determines the amount of WRM needed before desired capability is achieved.
- The number of bases available in-theater, the assurance of gaining access, and the quality of the bases also affect the ability to support aircraft in the theater.

Base Vulnerability

In selecting regions and locations for forward support locations, the vulnerability of the candidate locations to attacks from adversaries in future conflicts must be considered. Forward support locations could be primary targets for adversaries with long-range fixed-wing aircraft, cruise missiles, theater ballistic missiles (TBMs), or special operations forces, or primary targets of an attack by nonstate actors.[1]

Of these threats, theater ballistic missiles may be the easiest and least expensive for enemies to develop and deploy and the most difficult for the Air Force to defend against. The TBM threat is also the threat that is most sensitive to support location selection (because of the limited range of the majority of the world's ballistic missiles). Short- and medium-range missiles are the greatest threat to FSLs. Short-range (less than 600 nautical miles [nmi]) ballistic missiles are the most plentiful of the missile threats; there are tens of thousands of short-range ballistic missiles around the world, they are produced by more than 15 different countries, and are openly sold through weapons dealers.

Medium-range (600 to 1,500 nmi) ballistic missiles are less common than short-range missiles. Intermediate-range (1,500 to 2,500 nmi) and intercontinental (greater than 2,500 nmi) ballistic missiles are very expensive and a relatively small number of countries own them. In our vulnerability assessment, we therefore will focus on the short- and medium-range ballistic missile threat.

For a scenario in the Near East, most locations in Southwest Asia would be within reach of Iranian TBMs, while some locations in Turkey and all locations in the Eastern Balkans would remain out of reach.[2] In a Pacific scenario involving China, almost no location is safe from medium-range ballistic missiles, with the exceptions of

---

[1] The 1996 Khobar Towers and 2000 *USS Cole* attacks are two high-profile examples of attacks by nonstate actors.

[2] For most scenarios that might involve United States Air Forces in Europe (USAFE) or U.S. Central Command Air Forces (CENTAF), Diego Garcia, Northern Europe, and the United Kingdom are certainly the safest locations with respect to the TBM threat.

Guam, Australia, and some parts of Japan. However, many of the potential forward support locations in these regions are outside the range of most short-range missiles. Locations in Central Asia and the Indian Subcontinent are likely at risk from medium-range missiles; European, South American, and African locations are outside the range of most short-range missiles.

Base Access

The Air Force is confronted with the daunting challenge of securing base access in every conflict or operation. In general, the U.S. military has had an excellent record of maintaining working relationships with other host nations, which has contributed to many military successes in recent years. However, these relationships vary greatly, and in our assessment of current and potential forward support locations we must also evaluate the possibility of denial of access and its effects on combat capability, as was demonstrated during Operation Iraqi Freedom.

Arguably, one of the most important regions for potential forward support locations is Europe. European countries have been host to U.S. forces for more than 50 years. European forward bases have been used not only for local conflicts but also for operations in the Near East and Africa. The rich infrastructure, modern economies, stable democracies, and historical and cultural ties to the United States have made Europe an obvious choice for forward support and operating locations. Although there has been some political discontent regarding the resistance of France and Germany to support Operation Iraqi Freedom, such disagreements have by no means lessened the importance of European nations as hosts to U.S. forces.[3]

---

[3] Such disagreements, though disconcerting, should be expected even from the closest U.S. allies, as was demonstrated by the resistance of the United Kingdom, Spain, Italy, Greece, and Turkey to allow even over-flight rights during operation Nickel Grass, the airlift to Israel during the 1974 Israeli-Arab conflict (see Shlapak et al., 2002).

NATO has been expanded to include many of the former Soviet Bloc countries of Eastern and Central Europe. In 1999, NATO admitted Poland, Hungary, and the Czech Republic, and since then has admitted Bulgaria, Estonia, Latvia, Lithuania, Romania, Slovakia, and Slovenia.[4] The new NATO members and other aspirant countries have started programs with varying degrees of military relationships with the United States. Many of these countries played key supporting roles in OIF as well as in OEF and OAF (Tripp et al., 2004; Lynch et al., 2005). Romania and Bulgaria are of particular interest in this study because they are situated in proximity to regions of potential conflicts and have shown great interest in supporting U.S. forces in recent conflicts.[5] Romania and Bulgaria have significantly increased their defense spending to finance the radical restructuring of their militaries, and Bulgaria has also "adopted" the European and Euro-Atlantic defense and security values and considers its national security to be directly linked with regional and European security.[6] Both of these countries have several airfields suitable to support various strategic aircraft. Romania, for example, may have up to four airfields capable of supporting C-5s.

The United States continues to maintain a strong and sizable presence in Asia. Bilateral defense agreements with South Korea, Japan, Australia, Thailand, and the Philippines, along with other security commitments to some of the islands in the Pacific, ensure a continued presence of U.S. forces in the region. However, the bulk of U.S. forces are based primarily in South Korea and Japan in support of deliberate plans for that region. These forces are situated well for

---

[4] The current 26 NATO members along with 20 other nations belong to the Euro-Atlantic Partnership Council (see http://www.nato.int).

[5] Bulgaria allowed overflight rights during OAF despite domestic opposition. This was in contrast to Greece, a NATO member, which refused access to its airspace or airfields (Shlapak et al., 2002).

[6] Bulgaria increased its defense spending from $360 million in 2001 to $431 million in 2002. It also showed some interest in acquiring 22 F-16s from the United States but, as a result of limited funds, opted to upgrade its MiG-29s (see ISS, 2002). The statement about adoption of European defense values was made by General Nikola Kolev, Chief of General Staff of the Bulgarian Army, as reported in General Staff of the Bulgarian Army, 2002.

their primary mission in Korea, but their bases are remote from the Taiwan Strait and the South China Sea, where they may be needed for future regional conflicts. Guam is a valuable, well-developed U.S. territory in the Pacific, but the island is geographically distant from most potential conflict locations.

The U.S. Air Force keeps a small component in Singapore and in Australia and regularly holds military exercises with the Thai military. Nevertheless, many countries in the region may be wary of openly supporting a large, permanent U.S. presence in their territory, and others may not want to increase tension by taking sides in a potential conflict in which the United States, for example, aids Taiwan against the People's Republic of China. Therefore, regarding this region, we are concentrating both on potential sites for more permanent U.S. basing options (such as Darwin, Australia) and on virtual military bases or en-route support locations (such as U-Tapao, Thailand).

Bases—both temporary and permanent—in Australia, Thailand, Singapore, Malaysia, and the Philippines would greatly enhance the USAF combat support capabilities in support of a conflict in the Taiwan Strait, or operations against terrorism or insurgencies in Indonesia, the Philippines, or other critical regions in the Pacific Rim.

One of the most important regions in terms of security is the Near East, yet this region may be the most problematic for base access. The U.S. military kept a sizable presence in Saudi Arabia after Operation Desert Storm, but that decade-long arrangement was fraught with political and social issues.[7] After OIF, DoD decided to withdraw its troops from the kingdom. The United States has been successful in negotiating formal defense arrangements with Kuwait, Bahrain, Qatar, Oman, and the United Arab Emirates. However, as in the Asia-Pacific region, the granting of base and facility access does not necessarily mean guaranteed access. This was clearly evident in the reluctance of some countries in the region to support U.S. forces openly in OIF. A "democratic" Iraq may provide for an improved

---

[7] Whether justified or not, Osama bin Laden used the U.S. presence in Saudi Arabia as a rallying cry among the extremists in the region.

bilateral agreement in the future. However, a large and visible permanent presence by the United States may, once again, be used by extremists to undermine and limit access to resources in the region.

Given the large number of recent U.S. military operations and exercises in Africa, it may be of interest to consider potential facility locations on that continent. In 2004, the senior U.S. military commander for the European Command, General Charles Wald, stated that Sao Tome is a location of particular interest. The Voice of America (May 6, 2004) said that General Wald views the small African island state as a

> potentially ideal site for one of the U.S. military's so-called Forward Operating Locations. These are not permanent bases, but rather facilities that can be used in an emergency. General Wald said that, both because of its proximity to the oil-rich Gulf of Guinea and its strategic position along the Equator close to West and Central Africa, Sao Tome is an attractive location. He likened it to the Indian Ocean island of Diego Garcia, a strategically-placed base, used heavily by U.S. forces during the conflicts in Afghanistan and Iraq.

FSL Capability and Capacity

One of the major factors in selecting a forward support location is its transport capability and capacity. The parking space, the runway length and width, the fueling capability, and the capacity to load and offload equipment are all important factors in selecting an airfield to support an expeditionary operation.[8] Runway length and width are key planning factors and are commonly used as first criteria in assessing whether an airfield can be selected.

Table 4.1 outlines the airfield restrictions for some of the aircraft of interest.[9] The aircraft classification number (ACN) values relate aircraft characteristics to a runway's load-bearing capability, expressed

---

[8] In our analysis, some of these factors are computed parametrically in order to assess a minimum requirement of a potential field in order to meet a certain capability.

[9] For more information on airlifters and refuelers, see Appendix A.

as the pavement classification number (PCN). An aircraft with an ACN equal to or less than the reported PCN can operate on the pavement subject to any limitation on the tire pressure (U.S. Air Force, 2003). Each aircraft has a specified load capacity number (LCN) that identifies how much stress it is expected to exert on the runway.

All these factors combined dictate the type of aircraft that can be used at a base and the load capacity it can handle. The selection of each FSL will be based heavily on the airfield restriction.

Airlift

The time it takes to deploy personnel and equipment to a FOL is a decreasing function of the number of available aircraft. As an example, we consider the deployment of 3,000 short tons of materiel, roughly the equivalent of one each of Harvest Falcon Housekeeping, Industrial Operations, and Initial Flight line sets, to an operating location at a distance of 1,600 nmi.

The time required to deliver combat support materiel is essential in an expeditionary operation because the conflict or the humanitarian operation may be slowed or halted by delayed combat support resources. However, as illustrated in Figure 4.1, an increase in the aircraft fleet size is not the only means for reducing deployment time.

Table 4.1
Aircraft Airfield Restrictions

Aircraft Type	Minimum Runway Landing[a]		Minimum Taxiway (ft)	ACN (rigid pavement)		ACN (flexible pavement)		Ramp Space (ft^2)
	Length	Width		High	Low	High	Low	
C-130	3,000	60	30	8–34	11–41	6–30	8–34	15,519
C-17	3,500	90	50	22–52	22–52	18–52	22–71	47,500
C-5	6,000	147	75	8–29	11–39	10–37	17–54	62,724
C-141	6,000	98	50	16–48	21–68	17–49	21–70	31,362
KC-10	7,000	148	75	12–48	15–68	14–58	21–75	34,800

[a] Minimum runway distance required for landing with a full load (maximum takeoff weight).

Figure 4.1
Deployment Time as a Function of the Number of Aircraft

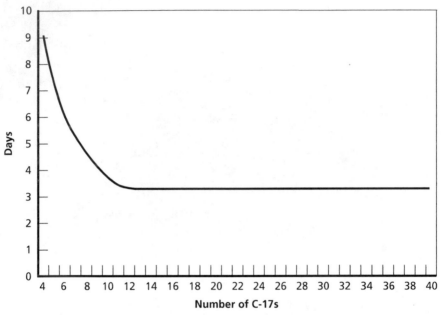

RAND *MG421-4.1*

As the number of aircraft increases, the deployment time decreases—but at a diminishing rate. At some point on the curve, the addition of more C-17s does not decrease the deployment time because throughput then becomes the limiting constraint. In the next section, we discuss the effect of throughput capacity on the deployment timeline.

Airfield Throughput Capacity
Another important factor in assessing the capacity of an airfield is the maximum-on-ground (MOG) capability.[10] MOG generally refers to

---

[10] Parking space, maintenance capacity, and the ramp space for storing and assembling the support equipment at an airbase are typically referred to as MOG (see Stucker et al., 1998). MOG and other factors determine the throughput of a base. In this report, we use the number of aircraft that can land, unload, be serviced, and take off per hour as a more effective measure of throughput constraint.

the maximum number of parking spaces an airfield can provide (parking MOG), but it can be specialized to include the maximum number of aircraft that can be served by maintenance, aerial port, or other facilities (working MOG). MOG can also refer to the maximum number of aircraft that can be refueled simultaneously (fuel MOG). In our analysis, we used both working MOG and parking MOG to compute the airfield capability or throughput with the following equation:

$$Throughput = \frac{MOG \times WorkDay}{ServiceTime} \quad ,$$

where *MOG* is the smaller of the working or parking MOG numbers, *WorkDay* is the number of working hours in a day, and *ServiceTime* is the required hours to load, unload, and service a particular aircraft.[11] Thus, *Throughput* is the maximum number of aircraft that can be processed through an airfield in one working day.

The diminishing return illustrated in Figure 4.1 is a result of an airfield's throughput capacity. An airfield may have a relatively large parking MOG but a small working MOG, reflecting both parking spots available for aircraft to be processed and the availability of the personnel and equipment necessary to process the aircraft. These constraints hold at both the destination (i.e., FOL) and at the originating airfield (i.e., FSL).

The smaller of the two MOGs will be the limiting factor. In the example above, we assumed that the constraining MOG was equivalent to two C-17s. Assuming 24-hour operations and 2.25-hour ground time, this configuration corresponds to a maximum airfield throughput of (24 × 2)/2.25, or just over 21 C-17s per day. Figure 4.2 presents the same deployment as in the previous example (3,000 short tons of materiel over a distance of 1,600 nautical miles) as a function of the number of C-17s, for various levels of working

---

[11] Lack of access to fuel MOG data prevented us from incorporating the fuel MOG in the equation.

Figure 4.2
Deployment Time as a Function of Airlift and MOG

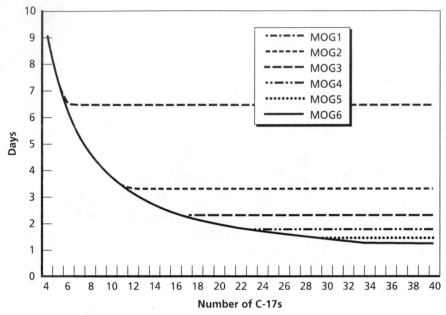

NOTE: MOG = maximum-on-ground.
RAND MG421-4.2

MOG. The graph shows six somewhat overlapping curves, each representing the deployment-time-versus-airlift tradeoff for a given MOG. As the number of airlifters increases, the corresponding decrease in deployment time occurs at a diminishing rate, as the curves in the figure show, until finally leveling off. This leveling off comes at a different point for each curve. For MOG 1, this point is reached at about six C-17s, and deployment time levels off at about 6.5 days. For MOG 6, deployment is reduced to 1.2 days using 34 aircraft.

Distance to FOL

As shown above, assigning greater numbers of aircraft alone may not reduce deployment time because of the constraining nature of a second factor of airfield MOG. Thus, investment in the infrastructure, personnel, and equipment at the airfields may also be required. A third factor is the deployment distance. For example, consider what happens when an aircraft must fly a longer distance for a deployment. If multiple sorties are required by a single airlifter, the longer distance is compounded by the repeated round trips, so that the aircraft makes fewer round trips per unit time than it would if the distance were shorter. The additional flying time per sortie, multiplied by the number of sorties necessary, gives the total increase in deployment time.

Figure 4.3 shows deployment time for 3,000 tons of materiel, as a function of the number of C-17 aircraft conducting the deployment, for various flying distances, under an assumed MOG of two. It shows that, as the number of airlift aircraft increases, the difference in deployment time caused by distance becomes less pronounced. For example, with five C-17s, deploying a distance of 500 miles takes 5.2 days, whereas deploying a distance of 1,500 miles takes 7.5 days, and

Figure 4.3
Deployment Time as a Function of Flying Distance

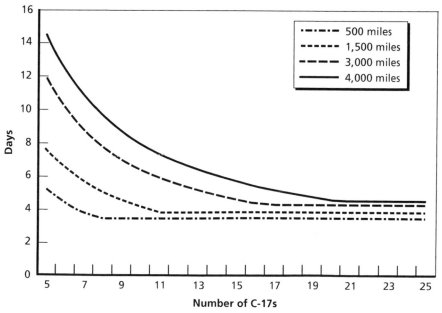

deploying a distance of 3,000 miles takes 11.8 days. With 10 C-17s, the airlift pipeline for the 500-mile deployment becomes saturated with aircraft, and deployment time levels off at 3.5 days. Assuming the same fleet size of 10 C-17s, the 1,500-mile deployment time also nearly levels off at four days and the 3,000-mile deployment requires 6.3 days. With 20 C-17s, the aircraft pipelines in all three deployments become saturated. In this case, a deployment distance of 500 miles takes, at a minimum, 3.5 days; 1,500 miles takes 3.7 days; and 3,000 miles takes 4.1 days.

As the figure shows, adding more airlifters to the system will reduce the deployment time, albeit at a diminishing rate, until the deployment time levels off due to MOG constraints. The figure also demonstrates that the point at which the system is saturated—that is, the point at which adding additional airlift aircraft will not decrease deployment time—varies as a function of the distance flown.

Thus, long flying distances affect deployment time most when airlifters are in short supply. With sufficient airlifters available, the effect of longer flying distances on deployment time can be minimal. Airfield throughput limitations appear to be the primary constraint on achieving more-rapid deployments.

Selection of Candidate FSL Sites

Beginning with the considerations mentioned above, we generated an initial list of over 300 potential FSL sites worldwide. Political restrictions, physical constraints, and data integrity were considered in an attempt to reduce this list to a more manageable size. We combined potential locations that were within the same country and were within 150 nmi of one another.

Current United States and United Kingdom Bases

The vulnerability of some overseas bases combined with potential limitations in accessing bases has highlighted the value of overseas territories of the United States and the United Kingdom. The United Kingdom has been a stalwart ally to the United States for many generations. For example, Britain enabled the 1986 raid on Libya. It was the only other country that shared the burden of enforcing no-fly zones in Iraq, and it supported the U.S. forces fully in Iraq despite the unpopularity of the conflict among the UK public.[12]

Some of the major U.S. bases outside of the continental United States are located in Guam and Alaska. These two locations, combined with bases in the United Kingdom and on the islands of Diego Garcia and Puerto Rico, can put most of the world within C-130 range of U.S. power projection capability (Shlapak et al., 2002). However, as was demonstrated in a previous RAND report (Amouzegar et al., 2004), the MOG constraint makes supporting even a moderately sized operation from this set of five FSLs impractical if speed of deployment and employment is a concern. Nevertheless, as evidenced by recent conflicts, the access afforded by bases such as Diego Garcia and Guam makes them invaluable in any future operation.

In fact, we might use Diego Garcia as a blueprint for future "acquisition" of readily accessible bases within other foreign territories. Countries such as the Philippines or Indonesia may, under certain circumstances, be willing to allow a long-term lease of some of their isolated islands for "permanent" use by U.S. armed forces, although significant infrastructure investments would be needed to bring the installation up to the standards required by U.S. forces.[13]

Afloat Prepositioning

Beyond the land-based sites, we also wished to examine the potential for storing WRM and munitions aboard an afloat preposition fleet (APF). The USAF currently leases a number of ships used for munitions storage. These ships are used to augment the munitions delivery capability of the air- and land-based munitions supply chain. Recent research conducted by the Air Force Logistics Management Agency

---

[12] However, Turkey, Saudi Arabia, and some of the Persian Gulf states allowed the use of their bases for Operation Northern Watch.

[13] Arguably, circumstances that would compel a government to "cede" sovereignty of a portion of its territory are very rare. Nevertheless, the possibility that such an opportunity may arise should not be discounted.

(Groothuis et al., 2003) has suggested some value in considering afloat preposition for the forward storage of nonmunitions WRM. We included both munitions and nonmunitions WRM afloat options, to compare their costs and benefits relative to land-based storage. Although afloat prepositioning does offer additional flexibility and reduced vulnerability versus land-based storage, the APF is much more expensive than land-based storage, and has serious risk with regard to deployment time. Even assuming a generous advance warning to allow for steaming toward a deployment's geographic region, it may be difficult to find a port that is capable of handling these large cargo ships. The requirements placed on the port, including preemption of other cargo movement, also restrict the available ports that can be used by an APF. Once the APF reaches its port, additional delays are incurred due to transferring the cargo to surface transport (usually trucks or rail). In this study, we assumed that all APF cargoes must be transported via surface from the port (which varies, according to the APF ship's "home" location and the deployment's geography) to the FOLs. Additional transferals to allow shipping via air or via other sea options were not allowed for the APF. We also assumed that all APF ships would have a seven-day warning time before the start of any deployment to allow for steaming toward the port associated with that deployment.

Alternative Modes of Transportation

Although most of the discussion in this chapter has focused on air transportation, ground transportation and sealift play a major role in the forward support basing architecture. Moreover, constraining factors—such as throughput, fleet size, and load capacity—apply to all modes of transportation and must be taken into consideration when there is an option to select an alternative mode.

There are several advantages to using sealift or ground transportation in place of, or in addition to, airlift. Ships have a higher hauling capacity than any aircraft and can easily carry outsized or superheavy equipment. Any water beyond twelve miles from the shore is considered international waters and thus can be navigated freely. Finally, ships do not require overflight rights from any foreign government.

Trucks are cheaper than aircraft or ships, of course, and are readily available in most locations through local contractors. They do not require specialized airfields and, although they are much slower than aircraft, under certain circumstances could contribute greatly to the delivery of materiel, especially when they are used in conjunction with airlifts. For example, with only 200 trucks, 10 Harvest Falcon sets (or about 11,000 pallets) can be delivered to various locations in SWA within 75 days. The same amount of materiel can be delivered in about 58 days using 24 C-17s, or in 85 days using 47 C-130s. The best single-mode result, 40 days, is attained using approximately 400 trucks. However, a mixed strategy of 200 trucks and seven C-17s can achieve the same goal in only 12 days.

Similarly, ships are slow relative to airplanes and may require specialized ports and equipment for loading and offloading. Nevertheless, sealift can be an effective alternative to airlift. For example, in a notional 4,000-nmi scenario comparing C-17s and the new large medium-speed roll-on/roll-off (RO/RO) ships, assuming no prepositioned ships in the theater, airlift could deliver only 72,000 tons of cargo in 36 days, whereas sealift could deliver 3,960,000 tons in the same number of days (Global Security.org, 2005). RAND has estimated that it would take about 13 days to deploy a Stryker Brigade package (about 16,000 tons and 4,500 personnel) from Ft. Polk to Europe using 60 C-17 equivalents at a throughput of two C-17s per hour, or using only two fast sealift ships at an average speed of 27 knots.[14]

One of the modes of transportation of special interest is the Fast Sealift Ship (FSS), which is used by the Military Sealift Command (MSC). These ships are RO/RO, with a range of about 12,000 nauti-

---

[14] See Vick et al. (2002) and Peltz, Halliday, and Bower (2003) for a fuller discussion of RAND analysis of Stryker Brigade closure times. The second work also includes a discussion of other timesaving alternatives, such as the prepositioning of Stryker units and materiel forward.

cal miles. The U.S. Navy owns eight FSSs, which are normally kept on reduced operating status but can fully activate and be under way to load ports within 96 hours. One of the main requirements for this type of ship is a harbor that can accommodate a large MSC vessel. In the absence of an adequate harbor, a roll-on/roll-off discharge facility (RRDF) and lighterage (to bring the cargo ashore) are needed. The RRDF is a floating pier that is set up to receive cargo from vessels off the coast of a deployment location. The cargo is offloaded to the RRDF and then loaded to a smaller sea vessel for transportation to the shore.[15]

An alternative to FSS is the Navy's fast combat support ship, a high-speed vessel designed as oiler, ammunition, and supply ship. This ship has the speed to keep up with the carrier battle groups. It rapidly replenishes Navy task forces and can carry more than 177,000 barrels of oil, 2,150 tons of ammunition, 500 tons of dry stores, and 250 tons of refrigerated stores (U.S. Navy Military Sealift Command, 2002).

A particularly attractive option includes high-speed sealift, such as use of the 91-meter wave piercing ferry International Catamaran (INCAT) 046 and the Revolution-120, a 120-meter wave piercing Catamaran.[16]

The HSS combines three attributes: light weight, high performance, and large payload. The INCAT 046 "Devil Cat," with a surface-piercing catamaran hull 91 meters long and a beam of 23 meters, is capable of carrying 500 metric tons and reaching speeds up to 43 knots.[17] The U.S. Army, as part of the Center for the Commercial Deployment of Transportation Technologies (CCDoTT) High-Speed Sealift program and in cooperation with the United States

---

[15] For a detailed description of RRDF operation, see Vick et al. (2002) and U.S. Navy Military Sealift Command (2002).

[16] These ships are manufactured by International Catamaran in its Australian shipyard.

[17] In 2002, a U.S. Army report examining a 600-ton capacity HSS gave estimates of $9.1 million annual lease cost and $8.8 million annual operating cost. See http://chl.wes.army.mil/research/projects/ribs/JLOTS%20Symposium/symposium%Wednesday/Schoenig/TSV_TSA_RDSymposium.ppt.

Transportation Command (USTRANSCOM) and the Maritime Administration, sponsored an evaluation of the 91-meter INCAT (Dipper, 1998). The newest INCAT design, Revolution 120, has turbine-powered jets and is 120 meters long with a beam of 30 meters. It can achieve speeds of more than 60 knots lightship (400 metric tons) and 50 knots fully loaded (1,200 metric tons).

The Royal Australian Navy used an INCAT-built catamaran, *HMS Jervis Bay*, to carry troops and vehicles to and from East Timor, a 430-mile run. It made up to three runs per week between Darwin, Australia, and Dili, Indonesia. According to the commander of the *Jervis Bay*, the catamaran was a definite advantage, given the lack of a port or port service in Dili.[18] The U.S. Marine Corps also used such catamarans in their tsunami-relief efforts.

The potential for alternative modes of transportation may play a major role in the selection of new forward support locations. The transportation options and combat support factors are both constraints and resources that we incorporate in the analytic framework discussed in the next chapter. Some of these factors are fixed (e.g., the location of a particular site); others are parameters (e.g., throughput capacity) that may be changed to examine the cost and benefit of additional investment to improve the capability of a FSL. There are, of course, other constraints, such as the political implications of regional imbalance, that should be considered but are beyond the scope of this study.

List of Potential Forward Support Locations

Ultimately, we settled on 50 potential sites, including Afloat Preposition Ships (APS), that we felt were sufficient for conducting the study. The final list is presented in Table 4.2.

---

[18] William Polson, "Navy Goes Down Under, Explores Future of Amphibious Warfare," http://www.c7f.navy.milnews/2000/09/16/html, September 16, 2000, cited in Vick et al. (2002).

Table 4.2
Potential Forward Support Locations

Bagram, Afghanistan	Tocumen IAP, Panama
Darwin, Australia	Clark APT, Philippines
Baku, Azerbaijan	Okecie, Poland
Shaikh Isa, Bahrain	Roosevelt Roads, Puerto Rico
Burgas, Bulgaria	Constanta, Romania
Djibouti Ambouli, Djibouti	Al Udeid AB, Qatar
Cotipaxi, Ecuador	Sao Tome/Salazar, Sao Tome
Beni Suef, Egypt	Dakar, Senegal
Ramstein AB, Germany	Paya Lebar, Singapore
Souda Bay, Greece	Louis Botha, South Africa
Andersen AFB, Guam	Moron AB, Spain
Chennai, India	U-Tapao, Thailand
Chhatrapati Shivaji IAP, India	Incirlik AB, Turkey
Balad, Iraq	Diego Garcia, UK
Aviano AB, Italy	Mildenhall and Welford, UK
Sigonella and Camp Darby, Italy	Eielson AFB, Alaska, US
Kadena AB, Japan	Hickam AFB, Hawaii, US
Misawa AB, Japan	APS Munitions 1, Diego Garcia
Yokota AB, Japan	APS WRM 1, Diego Garcia
Bishkek-Manas, Kyrgyzstan	APS Munitions 2, Guam
Kaduna Airport, Nigeria	APS WRM2, Guam
Masirah Island, Oman	APS Munitions 3, Mediterranean
Seeb, Oman	APS WRM 3, Mediterranean
Thumrait, Oman	APS Munitions 4, Okinawa
Masroor, Pakistan	APS WRM 4, Okinawa

Note: IAP = International Airport; APT = Airport.

CHAPTER FIVE

Analysis Methodology

We have developed an analytic framework to evaluate the effectiveness of alternative forward support basing architectures. These tools are designed to evaluate alternative forward support location configurations in an effort to identify combinations of FSLs that would perform well with respect to various measures of interest—such as facility and operating costs, deployment time, and transportation requirements—across a broad range of potential scenarios. The result is a collection of tools that may be used to answer questions ranging from the number and location of FSLs needed on a global scale to support contingencies around the world, to the optimal placement and transportation of materiel within a theater. Figure 5.1 describes our methodology for evaluating alternative FSL sites.

Our methodology begins with the selection of sample scenarios. These scenarios drive requirements for materiel such as base operating support equipment, vehicles, and munitions. We estimate these requirements using a RAND model, the *Strategic Tool for the Analysis of Required Transportation* (START) (Snyder and Mills, 2004). The START model builds requirements at the level of Unit Type Codes (UTCs) and, with the exception of munitions, does not estimate consumables (e.g., food and fuel).[1] The UTC is a natural unit to

---

[1] Some commodities (e.g., most general purpose vehicles) do not have a UTC or are commonly shipped as **Z99 UTCs (e.g., munitions). In these cases, each item is listed individually as a "**Z99" UTC. See Galway et al. (2002). Munitions are included because they

Figure 5.1
Overview of Our Analytic Process

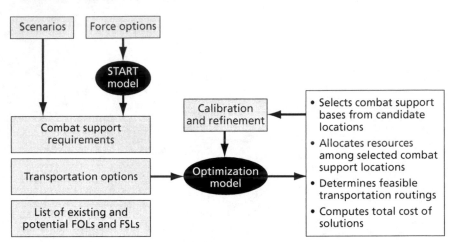

RAND *MG421-5.1*

quantify movement requirements because it forms the components of time-phased force deployment data sets. START combines the output list of UTCs with the Manpower and Equipment Force Packaging movement characteristics for each UTC. It converts the operational capability desired at a deployed location into a list of materiel and manpower needed to generate that capability.

These requirements, combined with the set of potential FSLs and FOLs that are derived from the scenarios, serve as the inputs to the optimization models that are central to this study. This set of deployments is then scheduled over a planning horizon to determine *timelines* that represent potential future deployment schedules. We then use another model, the *RAND Overseas Basing Optimization Tool* (ROBOT) (Amouzegar et al., 2004) to determine an optimal set of FSL locations for a given timeline, along with their inventory allocations, inventory requirements, transportation requirements, and the deployment timeline.[2]

---

require a considerable amount of airlift due to their weight, and unlike many consumables, cannot be procured on the local market.

[2] See Appendix B for a detailed description of the user's guide to the model.

This model is a Mixed Integer Programming (MIP) model, developed using the General Algebraic Modeling System (GAMS).[3] ROBOT explicitly models transportation constraints (e.g., number of transport vehicles, utilization rates, vehicle throughput), facility constraints (e.g., storage space constraints, net explosive weight [NEW] constraints), and time-phased demand for commodities at FOLs. The output from this optimization is the creation of a network that connects a set of disjoint FSL and FOL nodes. It allocates resources to a particular FSL and dictates the movement of combat support resources and munitions from FSLs to FOLs. The model also computes the type and the number of transportation vehicles required to move the materiel to the FOLs, as well as retrograde movements.

The ROBOT Model

We next present a simplified overview of ROBOT.[4] We begin by defining the following variables:

q_{jm} Number of mode m vehicles available at FSL j at the beginning of time 1

p_{jkmt} Number of mode m vehicles tasked to transport personnel,
u_{jkmt} p, munitions cargo, u, and nonmunitions cargo, y, from FSL
y_{jkmt} j to FOL k, beginning loading at time t

v_{jmt} Number of mode m vehicles available at FSL j at the end of time t

Constraints satisfying the limits on the total number of available vehicles system-wide, equal to the initial number of available vehicles C_m plus the variable r_m denoting the additional mode m vehicles pro-

---

[3] See Brooke et al. (1992).

[4] The complete mathematical model appears in Appendix C, with no explanatory discussion. For a complete mathematical programming formulation of the ROBOT model (including a detailed discussion) see Amouzegar et al. (2004).

cured, and the total vehicles available for loading at each FSL are defined as:

$$\sum_j q_{jm} \leq \left(C_m + r_m\right) \qquad \forall m$$

$$\sum_k [p_{jkmt} + u_{jkmt} + y_{jkmt}] \leq v_{jm(t-1)} \qquad \forall j, m; t \geq 2 \ .$$

FSL maximum-on-ground (MOG) constraints are defined in such a way as to account for both vehicle *space on the ground* and vehicle *ground time*. The MOG at each FSL is modeled separately for each of three *classes* of vehicles, because air, ground, and sea vehicles are assumed to use different loading equipment. Each FSL is assumed to have a maximum number of vehicle spaces allowed for loading for each class at any one time. Within a class of vehicles, different *modes* are assumed to consume differing fractions of this loading space. For example, the parking space for one C-5A is equal to the parking space for four C-130s (U.S. Air Force, 2003). Each of these differing modes of transport is also assumed to consume the loading space for a different length of time. Using the same example, the wartime planning loading time for a C-130 is 90 minutes; that of a C-5A is 255 minutes. Thus, each C-5A loaded will consume four times the loading space for nearly three times as long as will each C-130 that is loaded.

Consider one class (e.g., air) of vehicles, comprised of multiple modes (e.g., C-17, C-5), and let Aj be the MOG capacity for this class at FSL j. Then, defining α_m as the number of time periods necessary to load a mode m vehicle, the MOG constraints are defined over all modes m in the current class as:

$$\sum_{km} \sum_{g=0}^{\alpha_m - 1} \left(p_{jkm(t-g)} + u_{jkm(t-g)} + y_{jkm(t-g)}\right) \leq A_j \qquad \forall j, t \ .$$

The FOL maximum-on-ground constraints similarly restrict the FOLs based on the unload space available at each FOL.

Next, we define the variables:

w_j Binary variable indicating FSL j status, $w_j = 1$ if open, $w_j = 0$ otherwise

x_{ijkmt} Quantity of commodity i sent from FSL j to FOL k via mode m, beginning loading on time t

E_j Minimum units of storage needed for an economically feasible FSL at location j

n_j Additional square feet of storage space needed beyond E_j at FSL j

A demand constraint requires the cumulative arrivals by time t to satisfy at least a prespecified percent of the cumulative demand by time t. This constraint requires the declaration of parameter ω_{jkm}, equal to the number of time periods necessary to load a mode m vehicle at FSL j, transit to FOL k, and unload at FOL k. FSL storage constraints limit the space available for munitions and nonmunitions. The demand requirement and storage capacity are satisfied by the following constraints, respectively:

$$\sum_{jm,\,g \le t} x_{ijkm(g-\varpi_{jkm})} \ge D_{ikt} \qquad \forall i,k,t$$

$$\sum_{ikmt} x_{ijkmt} \le E_j w_j + n_j \qquad \forall j$$

$$n_j \le \left(F_j - E_j \right) w_j \qquad \forall j$$

where D_{ikt} is the cumulative demand, in tons, for commodity i at FOL k by time t, E_j is the minimum square footage needed for an economically feasible FSL at location j, and F_j is the maximum potential square feet of storage space at FSL j. Note that two versions of the storage space constraints exist for each potential FSL, one for munitions commodities and one for nonmunitions commodities, since separate storage is assumed for each. These constraints also control the opening and closing of FSLs.

A final necessary variable is:

z_{jkmt} Number of mode m vehicles tasked to make the return trip from FOL k to FSL j, departing at time t.

Once vehicles p, u, and y finish unloading at FOL k (assume that n represents the sum of the loading, transport, and unloading times), the following constraint reassigns those vehicles to return trips to FSLs:

$$\sum_j z_{jkmt} = \sum_j \left(p_{jkm(t-\varpi_{jkm})} + u_{jkm(t-\varpi_{jkm})} + y_{jkm(t-\varpi_{jkm})} \right) \qquad \forall k, m, t.$$

Note that this model formulation does not assign an individual transport vehicle to a single FSL, to a single FOL, or to a single commodity type. Instead, a given C-17 may transport munitions from FSL A to FOL B, and then make the return trip from FOL B to FSL C, where it will be loaded with a personnel cargo. Note also that individual FOLs are not necessarily "covered" by a single FSL. Instead, multiple FSLs may send commodities to a given FOL, if the optimal solution requires it.

The following constraint limits the average fleetwide utilization over the entire scenario duration to be less than the planning factor, σ_m, for each transport mode:

$$\sum_{jkt} \left(p_{jkmt} + u_{jkmt} + y_{jkmt} + z_{jkmt} \right) \le \sigma_m \left(C_m + r_m \right) \qquad \forall m.$$

The model is solved by finding a set of p_{jkmt}, q_{jm}, u_{jkmt}, v_{jmt}, w_j, x_{ijkmt}, y_{jkmt}, z_{jkmt} that (1) satisfies the set of contingency requirements and (2) minimizes the costs of conducting training and deterrent exercises over a given time horizon. That is, the peacetime costs of conducting training and deterrent missions are minimized while the solution set is constrained to have the storage and throughput required to meet "planned" contingency scenarios should deterrence fail. This is accomplished through scheduling major combat operations (MCOs) within our sequence of deployment timelines. The time-phased demands associated with these large contingencies ensure that the FSL network is capable of supporting these large demands. However, as discussed previously, the costs associated with these MCOs cannot be programmed. Thus, we include the constraints associated with the MCOs, but do not include their transportation costs in the optimization objective. Specifically, the formulation minimizes the net present value of opening and operating facilities, along with peacetime transportation costs, over a specific time horizon. ROBOT outputs a transportation plan and reports the time needed for FOLs to achieve initial and final operational capabilities. The ROBOT model can also be used to determine FSL postures that meet other objectives, such as minimal deployment time or minimal number of airlifters required.

A final caveat regarding our model needs to be made with respect to the input data provided. It should be apparent that the solutions returned would be sensitive to the set of scenarios provided. A vastly different set of input scenarios will likely return a different solution set of FSLs. This is why it is important to consider a broad range of potential future engagements in order to identify a robust set of facility locations. It is important to note that this model is not specifically tied to any one set of input scenarios, and changes to the inputs can easily be made if a given set of scenarios does not seem representative of some important consideration.

Portfolio Analysis of FSL Options

For any timeline, our optimization model can be used to identify an optimal FSL posture with respect to differing objectives, such as minimum cost, minimum deployment time, or minimum number of FSLs. Since the true optimal solution can only be determined if the future is known perfectly in advance, multiple timelines are considered individually, with the optimal FSL postures identified for each.

A portfolio analysis must then be undertaken to identify FSL postures that perform well across every timeline, to identify FSL postures that provide robust solutions.

Before the final FSL portfolio of options is generated, a refinement and calibration of the potential portfolios must be evaluated from a political point of view. The result may alter the FSL list and thereby affect the results of the optimization process. Some FSLs suggested by the model may be deemed impossible due to politics, practicality, or risk. Other FSLs not suggested by the model may merit consideration due to those same factors. These considerations may inform the inputs to another iteration of model runs. This postoptimality analysis can then iterate until an acceptable set of portfolios is determined.

The results of our analysis yield global portfolios of FSL structures and combat support materiel allocations. These portfolios include tables of metrics (such as policies, locations, technologies, and costs) that will allow policymakers to assess the merits of the various options. Ultimately, policymakers would thus be able to consider various mixes of FSLs, along with their respective capabilities and effectiveness, from a global perspective.

CHAPTER SIX

Selection of Overseas Combat Support Basing

In this chapter, we illustrate how the methodology developed in earlier chapters can be used in the selection of overseas combat support basing. We then present the alternative policy options available to the decisionmakers.

In our analysis, we posit that there will be small Air Force and Army forces permanently stationed in Europe, the Pacific (including Korea), and Southwest Asia. Because of the small number of forces permanently stationed in Europe, the Pacific, and SWA, our analyses deploy operational and support forces to exercise locations around the globe to conduct military exercises with allies and to ensure that forward operating bases can be used to support contingency operations should the need arise. The operational force packages are projected primarily from the CONUS, but the combat support resources will be provided from storage locations in or near the regions where the exercises take place. The model selects the best storage sites from the list of possible options. Finally, the storage facilities and the resources allocated to these sites are sized to support our global scenario, including potential major combat operations.

We will start the analysis by considering, as the baseline, the most likely global deterrent scenario, or *baseline scenario*. This scenario, presented in Figure 6.1, places the focus on supporting a number of deployments in the Persian Gulf region, Asian Littoral, and North Africa over a time horizon of six years, in keeping with the FYDP convention. The exercises vary in size of combat support requirements, as shown on the y-axis. The sizes of recent deployments

Figure 6.1
Baseline Scenario

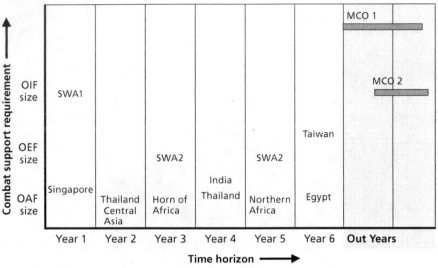

RAND *MG421-6.1*

are also given as a reference, including potential major combat operations.

We first evaluate the base scenario against existing U.S. overseas WRM storage locations (Ramstein AB, Germany/Sanem, Luxembourg; and Mildenhall, UK) and then expand the FSL list to include potential sites in other parts of the world.[1] The basic premise in each evaluation is to find the least-costly set of FSLs that has the capability and capacity to meet all the operational requirements, given all the constraining factors. These factors and other parameters are presented below:

- Cost
  - construction and/or expansion of facilities
  - operations and maintenance
  - transportation for peacetime and training missions

---

[1] Henceforth, we consider Ramstein AB as including Sanem as well.

- country cost factors
- Transportation options: land, sea, and air
    - land borders and canals
    - availability (~50 C-17s, Trucks, 4 HSS)
    - capacity in short tons (C-17: 45; Truck: 20; HSS: 400) or square feet (C-17: 1,080; Truck: 240; HSS: 40,000)
- Time-phased operational goals
    - 5 to 10 days IOC, 15 to 30 days FOC
- Storage capacity
    - weight and volume, net explosive weight (NEW)
    - afloat or land-based
    - throughput capabilities.[2]

Analysis of Existing Overseas Combat Support Bases

Figure 6.2 illustrates the geographical region of the baseline scenario as well as the locations of the existing overseas bases, including the munitions afloat preposition fleet. We solved the problem (i.e., we found the least-cost solution that satisfies the operational requirements), and the ROBOT model selected eleven FSLs from the set of existing locations (Table 6.1). It is interesting to note that the model selected neither the munitions afloat preposition ships nor any U.S. locations.

The transportation and operations and maintenance (O&M) costs, along with the construction cost, are presented in Figure 6.3. The construction cost includes upgrade and expansion to some of the existing sites to meet the operational demand required in the baseline

[2] A MOG of 2 was used for most of the FOLs. Every existing FSL had a MOG between 1 and 5, although some of the potential new FSLs had a substantially higher MOG (with one as large as 24).

Figure 6.2
Baseline Scenario Region and Existing FSLs

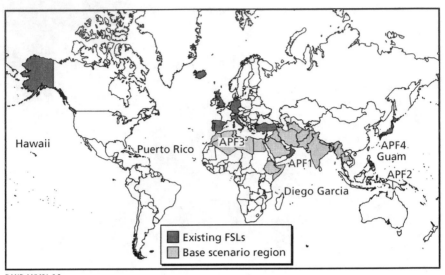

Table 6.1
Optimal Existing FSLs to Support the Baseline Scenario

Ramstein AB, Germany	Seeb, Oman
Sigonella AB and Camp Darby, Italy	Thumrait, Oman
RAF Mildenhall and Welford, UK	Kadena, Japan
Al Udeid AB, Qatar	Andersen AB, Guam[a]
Sheik Isa, Bahrain	Diego Garcia, UK
Masirah Island, Oman	

[a] The model did not select Andersen AB directly, mainly due to its remoteness and cost. However, the postoptimality analysis (Andersen is a "bomber island" with a large quantity of combat support resources) led to its inclusion.

scenario. It also includes additional munitions WRM igloos at Al-Udeid AB and Shaikh Isa, comprising 99,000 and 1,600 new square feet, respectively. Sigonella AB, Italy, would require about 121,000 square feet of additional storage space, and Diego Garcia requires a

Figure 6.3
Total Cost of Meeting the Baseline Scenario Requirements Using
Existing Bases

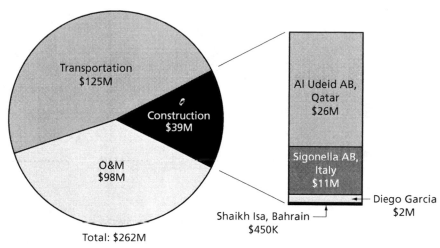

marginal expansion of an additional 11,000 square feet for nonmunitions WRM storage.

All modes of transportation were used in support of the baseline deployment. Trucks and HSS contributed to roughly half of the transportation capability (see Table 6.2).

When no infrastructure expansion was allowed, the APF munitions ship that was assigned to the Arabian Sea was included in the solution set (see Figure 6.4). Although the operational requirements

Table 6.2
Transportation of Assets for the Baseline Scenario
(percent of total)

Asset Type	C-17	HSS	Truck	Total
BEAR	15	7	3	25
Munitions	22	16	19	57
Rolling Stock	11	5	2	18
Total	48	28	24	100

Figure 6.4
Alternative Options Are Possible at Additional Cost

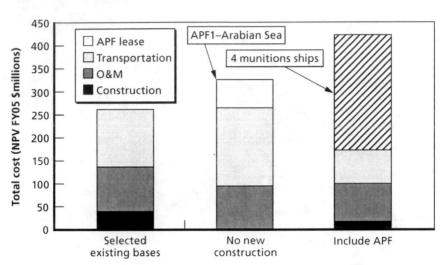

NOTE: NPV = net present value.
RAND MG421-6.4

were met, the overall cost was increased by about 25 percent (from $262 million to $327 million).

Figure 6.4 also presents the overall cost of $420M that is incurred if all four APF munitions ships are used in support of the baseline scenario. A large portion of the cost results from APF lease and maintenance, with some expansion and new construction at Sigonella, Diego Garcia, Seeb, and Shaikh Isa. This, of course, does not take into account the nonmonetary value of APF ships—such as access issues or risk mitigation. Ignoring other policy issues with respect to the munitions APF, the model shows that the most cost-effective solution comprises land-based storage facilities.

We next assess the capability of the existing bases against a set of differing deployment timelines (streams of reality). The details of these scenarios were outlined in Chapter One and the timelines were presented in Chapter Three. However, for convenience, we present again in Table 6.3 the sequencing of the scenarios.

Table 6.3
Sequencing of Scenarios by Timeline

Year	Base Scenario	Stream 1	Stream 2	Stream 3	Stream 4
1	SWA 1	SWA 3	SWA 1	South America 2	Spratleys
	Singapore	Southern Africa	Horn of Africa	Cameroon	Chad
		East Timor		Singapore	
2	Central Asia	Thailand	Central Asia	SWA 3	South America 1
	Thailand	Sierra Leone	Liberia	Thailand	Horn of Africa
				Haiti	
3	Horn of Africa	Spratleys	Balkans	Taiwan	SWA 2
	SWA 2	Haiti	Rwanda	S. Africa	Singapore
		Chad			
4	Thailand	Balkans	Singapore	Spratleys	Taiwan
	India	Egypt	Cameroon	Egypt	Haiti
			India		
5	SWA 2	SWA 1	SWA 2	SWA 1	SWA 2
	North Africa	North Africa	Taiwan	Rwanda	East Timor
		Liberia	Sierra Leone	East Timor	
6	Egypt	Central Asia	Spratleys	Central Asia	SWA 1
	Taiwan	India	Chad	North Africa	Rwanda
		Cameroon	Thailand	Singapore	
7+	MCO 1	MCO 1	MCO 1	MCO 1	MCO 1
	MCO 2	MCO 2	MCO 2	MCO 2	MCO 2

NOTE: SWA = Southwest Asia; MCO = major combat operation.

Capability of Existing Bases to Support Other Scenarios

For each new timeline, we seek a set of optimal locations and materiel assignments that meets the deployment requirement while minimizing the overall cost to the system. As before, the set of overseas storage locations must be able to meet the two-MCO requirement as well. We set all of the facility utilization values from the baseline solution as lower bounds for the subsequent Streams 1–4, to determine the additional facility requirements needed to meet these geographically dispersed sets of scenarios.

We continue to consider only the existing WRM and munitions storage locations. The minimized cost for all timelines is presented in Figure 6.5. Note that the construction cost for each stream remains relatively steady, indicating that little additional construction is required for Streams 1–4 beyond that of the baseline scenario solution. Although the O&M costs also do not vary greatly across timelines, note that the transportation cost is increased dramatically for Streams 3 and 4, which have a few deployments to far reaches of the globe. The selected existing FSLs that appear in the baseline scenario solution (see Table 6.1) continue to support Streams 1 and 2. However, because of the location of some of the deployments in Streams 3 and 4, a portion of the combat support assets had to be reallocated to an additional FSL in Eielson, Alaska. For Stream 4, the ten-day IOC

Figure 6.5
Total Cost of Supporting Each Stream

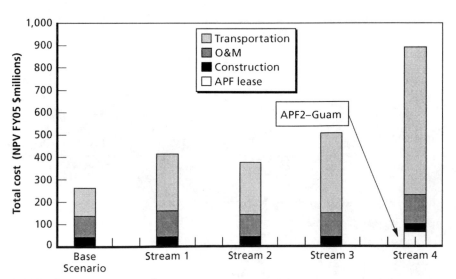

NOTE: NPV = net present value.
RAND MG421-6.5

requirement for the South American deployment could not be met from the existing locations. From this set of locations, IOC could not be met until 12 days and required the use of the APF munitions ship in Guam to pick up the slack in storage and transportation requirements.

Analysis of Potential Combat Support Overseas Bases

In Chapter One, we generated a list of potential FSL locations around the globe that could support a wide range of deployments. We started the analysis of this larger set of potential locations by evaluating the baseline case or the most likely timeline presented earlier. Figure 6.6 illustrates the location of the potential and existing FSL sites considered, along with the optimal set of FSLs selected by

Figure 6.6
Potential and Existing Locations for Combat Support Basing

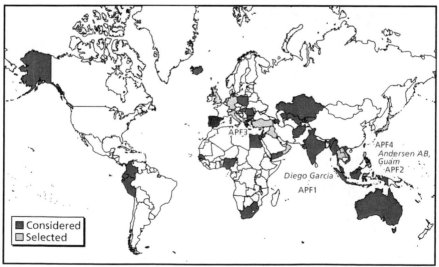

the model for the baseline scenario. The 11 existing sites presented earlier in Table 6.1 remain in the solution, along with five new sites in Europe and Asia. Note that the 11 existing sites were not "forced into" the solution here but were selected by the model as the most cost-effective of the 50 sites shown in Table 4.2. The selected FSLs are listed in Table 6.4. Note, however, that this list is by no means sacrosanct, and alternative sites may provide the same capability at a similar or marginally greater cost. In particular, Souda Bay, Greece; Akrotiri, Cyprus; Constanta, Romania; or Burgas, Bulgaria may be suitable alternatives to Incirlik, Turkey. We note such alternate sites whenever possible.

Figure 6.7 shows the savings in using the bases from the expanded set. Although the construction cost is greater than that for existing bases, the overall cost is reduced by about 30 percent. This is not surprising because the set of bases was selected to optimize the total cost. That is evident by the large reduction in the transportation cost.

Next, we consider the effect of APF munitions ships on the site selection. We did this by forcing the model to keep the APF munitions ships independent of their cost and utilizing them to the extent

Table 6.4
Optimal FSLs from an Expanded Set to Support the Baseline Scenario

Ramstein, Germany	Seeb, Oman
Sigonella and Camp Darby, Italy	Sheik Isa, Bahrain
Mildenhall and Welford, UK	Thumrait, Oman
Al Udeid AB, Qatar	Incirlik, Turkey
Masirah Island, Oman	Clark AB, Philippines
Andersen AB, Guam	Paya Lebar, Singapore
Diego Garcia	U-Tapao, Thailand
Kadena, Japan	Balad, Iraq

Figure 6.7
Total Cost of Using Existing Bases Versus the Expanded Set to Support the Baseline Scenario

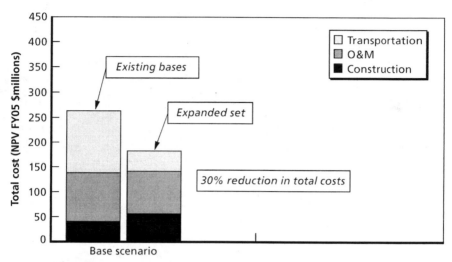

NOTE: NPV = net present value.
RAND *MG421-6.7*

possible. The result was the deselection of Balad (Iraq), Paya Lebar (Singapore), and two of the existing bases at Sigonella (Italy) and Shaikh Isa (Bahrain). Naturally, the overall cost is increased, although the cost is dominated by the leasing of the munitions APF ships. Figure 6.8 illustrates the results for three alternatives (existing bases, expanded set of bases, and the expanded set with four APF munitions ships). It also lists the land bases selected along with the four APF munitions ships.

Evaluation of Overseas Bases Against the Alternative Streams of Reality

The test, of course, is the capability of the selected FSLs across a series of deployments and scenarios. To ensure that the planning is robust, we examined the capability of existing and potential bases against the four streams of reality listed in Table 6.3. As noted earlier, each

Figure 6.8
Cost of Alternative Options to Support the Baseline Scenario

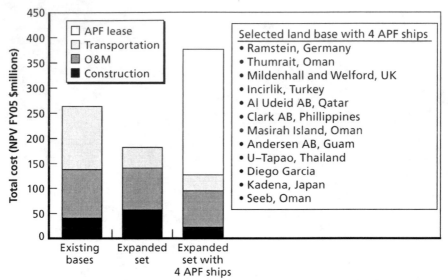

NOTE: NPV = net present value.
RAND *MG421-6.8*

timeline introduces additional complexity by expanding the geo-graphical net and increasing the combat support requirements.[3]

As was done for the analyses of existing storage locations only, after we obtain the baseline scenario solution for the expanded set of bases, we set all the facility utilization values as lower bounds for the facility utilizations of subsequent timelines, to determine the addi-tional facility requirements needed to meet the more stressing time-lines. For each stream, the solution to the baseline scenario did not have sufficient capability to support the increased demand, so addi-tional FSLs were required. Figure 6.9 illustrates the results for the first two streams: FSLs were required at Puerto Rico (to support Haiti deployment), Sao Tome/Salazar, Louis Botha (South Africa), and

---

[3] In comparison to the baseline, Stream 1 requires additional combat support of 2 percent. Streams 2 through 4 require additional combat support of 4 percent, 8 percent, and 12 per-cent, respectively.

Figure 6.9
Additional FSLs for Streams 1 and 2

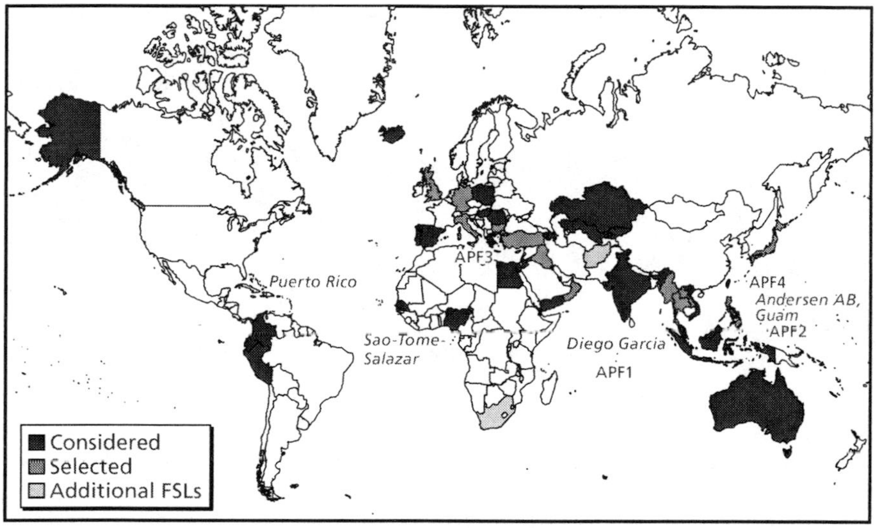

RAND *MG421-6.9*

Bagram (Afghanistan), in addition to the baseline FSL solution set
(see Table 6.4).[4]

The overall cost to support these two streams is greater in comparison to that of the baseline, which is not surprising because of
their increased combat support requirements (as Figure 6.11 will
show). Note that for each timeline the expanded pool (i.e., selection
from existing and potential FSLs) offers the least-cost option with the
same combat support capabilities when compared to the existing set
of locations. More important, it should be noted that including these
new sites (e.g., Sao Tome) offers increased flexibility and capability
(i.e., they can support a wider range of operations) with marginal increase in construction and O&M cost from the baseline solution set
of FSLs.

---

[4] An alternative site to Haiti may be Costa Rica. Note also that Sao Tome was the only additional FSL to appear in both Stream 1 and Stream 2 solutions.

The results for Streams 3 and 4 offer the same conclusion: (1) the optimal set of FSLs selected from the expanded pool offers the same capability at a lower cost in comparison with the existing bases, and (2) global deterrence requires a global strategy that can be achieved by a portfolio of overseas bases with marginal increases in construction and O&M cost from the baseline scenario. Figure 6.10 illustrates the location of these sites on the map and Table 6.5 lists the bases.

The map divides these locations into Tier 1 and Tier 2 categories. Both Kaduna, Nigeria, and Sao Tome appear in the solution for each of Streams 1–4. Due to the proximity of these two locations, potential FSL locations in the Gulf of Guinea should be examined closely. Louis Botha is quite remote from these West African locations, and may need to be examined on its own. The need for an increased South American presence is suggested by the fact that Roosevelt Roads, Tocumen, and Cotipaxi appear in the solution sets

Figure 6.10
Supporting Global Deterrence Using a Global Set of Overseas Bases

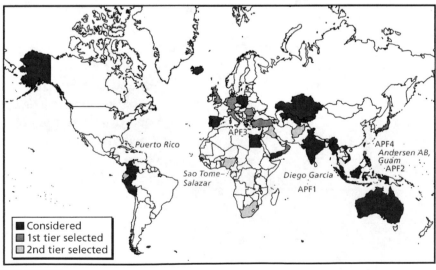

Table 6.5
Global Set of Overseas Bases

Tier 1	Tier 2
Al Udeid AB, Qatar	Louis Botha, South Africa
Andersen AB, Guam	Bagram, Afghanistan
Diego Garcia	Baku, Azerbaijan
Kadena, Japan	Roosevelt Roads, Puerto Rico
Masirah Island, Oman	Tocumen, Panama
Mildenhall and Welford, UK	Cotipaxi, Ecuador
Ramstein, Germany	Sao Tome/Salazar, Sao Tome
Seeb, Oman	Kaduna, Nigeria[b]
Sheik Isa, Bahrain	Balad, Iraq
Sigonella and Camp Darby, Italy	
Thumrait, Oman	
Clark AB, Philippines	
Incirlik, Turkey	
Paya Lebar, Singapore	
U-Tapao, Thailand	
Souda Bay, Greece[a]	

[a] Alternatives to Souda Bay, Greece, are Akrotiri, Cyprus; Burgas, Bulgaria; or Constanta, Romania.
[b] An alternative to Kaduna, Nigeria, may be Dakar, Senegal.

of Streams 3, 2, and 1, respectively. An increased forward presence in Central Asia is suggested by the fact that Bagram appears in the solution for 2 timelines, while Baku appears in one timeline solution. We use the label "Tier 2 FSLs" for these alternative sets of sites that deserve more detailed consideration as potential FSLs (Sao Tome/Kaduna; Louis Botha; Roosevelt Roads/Tocumen/Cotipaxi; Bagram/Baku). Additionally, all of these Tier 2 FSLs (with the exception of Puerto Rico) have uncertain political futures or limited internal capabilities. Iraq also falls in the "uncertain future" category, but its location for support of many operations makes it invaluable. However, as we have mentioned earlier, one should not focus on a particular latitude and longitude but rather on a particular region. Balad, Iraq, is suitable if all the issues of security and long-term political amenities are resolved. If the uncertainties continue, then an alternative location in the region with similar capabilities should be considered. Thus, we categorize Balad as a Tier 2 FSL also.

Figure 6.11 presents the costs for all the streams. Note especially that the set of existing land-based FSLs could not support Stream 4 requirements and required that the IOC deadline be extended from 10 to 12 days for the South American deployment, and also required the use of an APF munitions ship. However, when we selected from the expanded set of land-based FSLs, the need for the afloat option disappeared and the ten-day IOC deadline was met for the South American deployment.

The Effect of Global Basing on Lift

The savings from the expanded pool of FSLs are not limited to the total dollar reduction. They also include an efficient usage of the

Figure 6.11
Total Cost of Supporting Streams 1–4 Using Existing and Expanded Set of FSLs

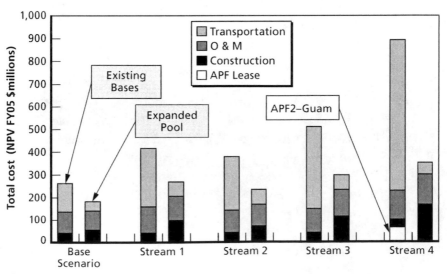

limited transportation assets. Figure 6.12 illustrates the allocation of combat support materiel to different modes of transportation. The left-hand bar for each timeline represents the solution using the existing bases and the right-hand-side bar represents the solution from the expanded pool of FSLs. It is clear that for each timeline the model was able to make better use of trucks and high-speed sealift for the expanded pool of bases, yielding about 50 percent less airlift usage without compromising operational requirements. Moreover, using the expanded pool of locations (or the APF, as in Stream 4 existing bases) can reduce total reliance on transport, due to collocation of storage sites at FOLs. This is indicated by the striped blocks in the figure.

A Portfolio of Options

For all five timelines considered (the base scenario and each of the four streams), we used our optimization model to identify an optimal

Figure 6.12
Transportation of Materiel Across the Streams as Portion of Total Lift

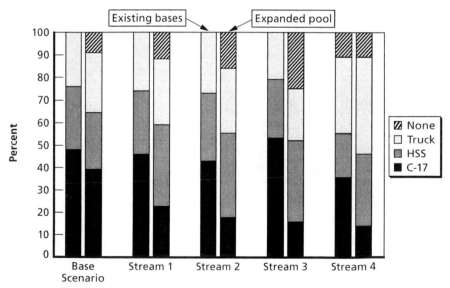

FSL posture with respect to a minimum cost objective. However, as discussed in Chapter Five, the truly optimal solution can be determined only if the future is known perfectly a priori. Therefore, a portfolio analysis is needed to identify FSL postures that provide robust solutions across every timeline. Hence, we evaluated the robustness of our *recommended set* of FSL locations, which includes all 16 Tier 1 locations appearing in Table 6.5, along with Tier 2 locations Balad, Louis Botha, Bagram, Sao Tome, and Cotipaxi. We sized each facility using the results of our earlier analysis and then tested the performance of this FSL posture against all five timelines. This set of locations was able to provide sufficient capability to meet all of the combat support requirements for each stream of reality (Stream 4 was an exception, requiring additional storage space at Cotipaxi) at a lower cost than could be achieved through the use of the existing bases. The minimized costs for the recommended set, along with the earlier results from the existing bases and the expanded pool, are presented in Figure 6.13. Note that the construction and O&M costs remain constant for the recommended set across all timelines (with the exception of Stream 4, which required additional infrastructure at Cotipaxi). Only the transportation costs vary.

An important distinction should be noted here. The costs for both the existing bases and the expanded pool of locations have been minimized independently of each timeline. That is, the model optimized the FSL posture for each given stream of specific deployments. However, for the recommended set, we have fixed the location and the capacity of each FSL and optimized across all the timelines. While the total cost of the recommended set is marginally higher for the baseline scenario (5 percent higher than the existing set of bases), it is considerably more capable than either of the other two baseline scenario solutions. Across Streams 1–3, the recommended set is on average 28 percent less costly than the various solutions for the existing set of facilities. Over these same three timelines, the recommended set provides solutions that are on average within 16 percent of the minimal cost solutions for the expanded pool. For Stream 4, the recommended set is able to meet the ten-day IOC deadline for the

Figure 6.13
Comparison of Total Cost for Supporting All Scenarios Using Existing, Expanded, and Recommended Sets of FSLs

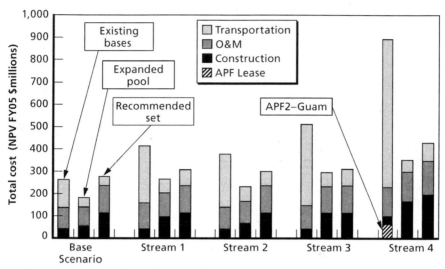

NOTE: NPV = net present value.
RAND *MG421-6.13*

South American deployment, with a cost that is 52 percent less than the cost of the existing set (which requires 12 days to meet IOC) and only 22 percent greater than the minimal cost solution for the expanded pool.

Potential for Alternative Solutions

We have emphasized throughout this report that our results should not be interpreted to suggest that a specific set of locations is the "best" solution. Rather than focusing on a specific site recommendation, these results suggest that an FSL storage capability should be pursued in particular geographic regions (e.g., a West African capability, instead of a single location in Sao Tome). To demonstrate the flexibility allowed to policymakers in this area, we tested two alternative recommended sets of FSLs, each with 17 locations (contrasted with the 21 locations appearing in the earlier recommended set to allow for more consolidation of resources). Table 6.6 presents these two sets of alternative locations.

Note that 12 of the 17 locations appear in both alternatives, signifying the importance of these locations. For the five remaining locations, we allowed the model to examine alternative regional options, such as Souda Bay versus Burgas or Dakar versus Kaduna. As with the previous recommended set, for each of the new sets we determined a facility capacity for each location and fixed that facility capacity for the computational testing across all five timelines. The minimized costs associated with the three recommended sets for each of the five timelines are presented in Figure 6.14. As noted before, the construction and O&M costs remain constant for each of these FSL postures across all scenarios (again with the exception of Stream 4, for which

Table 6.6
Alternative Recommended Sets of Overseas Bases

Alternative Recommended Set 1	Alternative Recommended Set 2
Al Udeid AB, Qatar	Al Udeid AB, Qatar
Andersen AB, Guam	Andersen AB, Guam
Diego Garcia	Diego Garcia
Kadena, Japan	Kadena, Japan
Masirah Island, Oman	Masirah Island, Oman
Mildenhall and Welford, UK	Mildenhall and Welford, UK
Ramstein, Germany	Ramstein, Germany
Seeb, Oman	Seeb, Oman
Sheik Isa, Bahrain	Sheik Isa, Bahrain
Sigonella and Camp Darby, Italy	Sigonella and Camp Darby, Italy
Thumrait, Oman	Thumrait, Oman
Clark AB, Philippines	Clark AB, Philippines
Paya Lebar, Singapore	U-Tapao, Thailand
Souda Bay, Greece	Burgas, Bulgaria
Baku, Azerbaijan	Bagram, Afghanistan
Dakar, Senegal	Kaduna, Nigeria
Tocumen, Panama	Cotipaxi, Ecuador

Figure 6.14
Comparison of Total Cost for Supporting All Scenarios Across Three Alternative Recommended FSL Postures

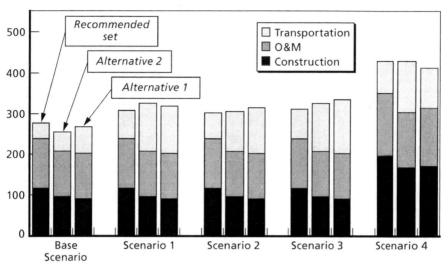

RAND *MG421-6.14*

additional infrastructure is needed at the South American storage sites).

The two alternative sets perform very similarly to the earlier recommended set: Across all five streams the average increase in expenditures is 1 percent and 2 percent for sets 1 and 2, respectively. Since we have already shown that the initial recommended set performs robustly across all five timelines, we are similarly satisfied with the performance of the alternative recommended sets. This demonstrates the flexibility available when examining potential sites for new FSL locations. The focus should be on geographic regions, and an attempt should be made to identify sites with nearby locations and similar capabilities (e.g., potential for multimodal transport).

Thus, any of the recommended sets of FSL locations and capacities provides a robust solution that performs well across a variety of timelines and deployments. Other solution sets could be identified that provide similar performance if other considerations (e.g., political or strategic) precluded the use of our recommended solutions.

CHAPTER SEVEN

Conclusions and Recommendations

The geopolitical divide that defined the Cold War era has ceased to exist and has been replaced with an international system that is fluid and unpredictable. The Air Force has responded to this changing security environment by transforming itself into a more expeditionary force. The ability of U.S. forces to provide rapid and tailored responses to a multitude of threats across the globe is a crucial component of security in today's political environment.

The Air Force is now fully committed to the Air and Space Expeditionary Force concept and the transformation needed to enable the Air Force to project power quickly to any region of the world. To support military operations seamlessly and efficiently in all phases of deployment, a global contingency support basing architecture must be designed.

This report has presented and explained an analytic framework that can be used to select a robust set of forward support locations from a list of potential, feasible candidates, for use in examining war reserve materiel storage options. Our analytic framework uses a combination of capability-based planning and threat-based scenarios to assess the effectiveness of alternative forward support basing architectures. At the center of our framework is a mixed integer optimization program, the *RAND Overseas Basing Optimization Tool* (ROBOT), which optimally allocates combat support resources to selected locations and at the same time develops an efficient transportation network. ROBOT minimizes total peacetime support costs while meeting the time-phased operational demand for a given set of constrained resources.

In keeping with the current security paradigm, and the Multi-Period-Multi-Scenario (MPMS) concept, we constructed several deployment scenarios (streams of reality, or timelines) using the following guidelines:

- The selection process should not be slaved to a particular deployment and thus should not optimize the system for a handful of deployments.
- Deployment scenarios should cast a wide geographical net in order to stress the combat support and transportation requirements.
- Deployments should be sequenced in time and space in order to evaluate physical reach and test long-term effect of location and allocation of assets.
- To hedge against uncertainty in the future security environment, a series of possible scenarios should be developed to test the robustness of the overseas combat support bases.

Each timeline considered several deployments (exercises, deterrent missions, etc.) over a six-year horizon. At the end of each timeline, two major combat operations (MCOs) were added to correctly size the storage capacity of each overseas combat support base. Using defense planning guidance and other resources, we started with a "most likely" timeline focusing mainly on the "hot regions" of the world. Other scenarios were added in order to stress the system and assess the robustness of the FSL selections. In keeping with the principles stated in the report, the model does not cost major combat operations even though constraints associated with these MCOs are included in the model.

Our analytic framework can address a wide range of basing decisions. We first examined only the existing WRM and munitions storage locations against a set of different timelines starting with the "most likely scenario" and allowed the model to select a set of FSLs from the existing set. We then selected a set of FSLs from a larger pool of FSLs that included both existing and potential combat support bases. The results of these computational tests were a set of FSLs that allowed for a 30–60 percent relative reduction in cost and up to a 50 percent reduction in airlift usage while continuing to meet the required deployment times. The results also support the concept that global deterrence requires a global strategy, which can be achieved by a portfolio of overseas bases with marginal increases in construction and O&M costs over the current system.

One of the key strengths of the analytic framework used is the depth at which each candidate FSL is considered. The vulnerability of each FSL to attacks from adversaries was examined when selecting potential regions and locations. The access to these potential sites should a conflict arise was also taken into account. While the United States has maintained excellent relationships with host nations, we still need to consider the possibility that we will not have access to some of these locations should the need arise to utilize them. Finally, we also needed to take into account the different constraints on transportation of materiel to and from the FSL location.

We therefore make the following recommendations to act on these conclusions:

Using a global approach to select combat support basing locations is more effective and efficient than allocating resources on a regional basis. One of the strengths of our analytic framework is the lack of regional command boundaries. We are able to look at all regions of the world simultaneously with operations occurring in various locations at the same time, thereby extracting the most efficient solution without adversely compromising the capability needs of a particular region. Currently, the Air Force lacks a focal point for managing its investment in global infrastructure. Combatant Commanders influence their assigned warfighting units, which in turn influence Air Force investments on a regional basis, but there is no central organization that has the overall responsibility to investigate how these regional capabilities interact to provide global force projection capabilities. One option to overcome this shortfall would be the creation of a centralized Air Force planning and assessment group at the Air Staff. Because the potential scenarios impacting U.S. interests are constantly shifting, such a group needs to continually revise the model inputs and rerun these computer models, to ensure that the logistics posture is well suited to the current environment. This group might also have the budgeting and Programmed Objectives Memorandum preparation responsibilities associated with global logistics infrastructure.

Political concerns must be addressed in any decision on potential overseas basing locations. For instance, although an afloat preposition fleet is much more expensive than alternative land-based storage options and may suffer from increased risk in deployment time, it may be necessary to consider the APF option because it offers more flexibility if access is denied. Additionally, a model that has cost and time as its major driving criteria continually selects such countries as Iraq. However, the uncertainty surrounding the future of Iraq (and similar countries) should force us to pause and consider other alternative sites that may be less desirable mathematically but offer higher probability of access and stability.

Closer attention should be paid to Africa both as a source of instability and as a possible location for combat support bases. With its potential as a source of future oil, combined with the uncertain future of many of its nation states, Africa requires a great deal of attention by policymakers. Northern and sub-Saharan Africa continue to be plagued by civil wars, ethnic or clan-based conflicts, and/or severe economic disasters. There is a greater likelihood that terrorists may seek haven in the remote areas of Africa because of the continued U.S. military presence in the Middle East and Southwest Asia. Given the possibility that deployments to the region will increase in the future, the current set of bases will not support these operations. Possible FSL locations in Africa could support operations across the entire southern half of the globe. Although the initial construction costs for these bases would be high, they would be quickly offset by the reductions in transportation costs. As an initial phase, we recommend that the western regions of Africa be closely evaluated, with particular attention to Nigeria, Sao Tome/Salazar, South Africa, and Senegal. The development of African FSLs could be tied into other foreign policy and outreach initiatives in Africa, such as the NATO Mediterranean

Dialogue country relationships with Algeria, Mauritania, Morocco, and Tunisia.

Some of the Eastern European nations should be considered as serious candidates for future overseas bases. The potential for continued conflicts in Central Asia and the Near East has made many of the countries in the eastern part of Europe very attractive as potential storage locations for WRM. The appeal of this region has been further heightened by the inclusion of some of these countries in the EU and NATO, their lower cost of living, and their relatively high-quality professional labor market. Romania and Bulgaria in Eastern Europe, along with Mediterranean locations such as Greece and Cyprus, would allow easy access to both CENTCOM and EUCOM. These locations are especially attractive because they allow for multi-modal transport options using Black Sea ports for Romania and Bulgaria (assuming passage through the Bosporus Strait to the Mediterranean). Poland and the Czech Republic, though very accommodating to U.S. efforts in the current operations, are located relatively far from the potential deployments considered in this report. The Czech Republic is a landlocked state, and—although Poland has a significant coastline on the Baltic Sea—its ports do not allow for rapid transport to the regions of interest to the U.S. Air Force. In terms of transportation time and cost, neither Poland nor the Czech Republic offers savings versus the existing installations in Germany, and either would require a substantial investment in transportation infrastructure to attain the current capability levels in Germany.

Southeast Asia offers several robust options for allocation of combat support resources. The remoteness of Guam and Diego Garcia from most potential conflicts in the region requires the consideration of other locations in the Pacific. The geographical characteristics of the U.S. Pacific Command (PACOM) puts a heavy reliance on airlift and possibly fast sealift. Most of the current U.S. bases are located in Japan and the Korean Peninsula with the main purpose of supporting the Korean deliberate plan. To support other possible contingencies, we propose a closer examination of three locations: Thailand, Singapore, and the Philippines. Each of these locations offers a host of options for the Air Force, including storage space, adequate runway facilities, proximity to ports, and strategic location. Darwin, Australia, has many of the desired attributes for an overseas combat support base, but its remoteness to most potential conflicts makes it a relatively poor choice.

Potential future operations in South America may be greatly constrained unless additional infrastructure in the region is obtained. In our analysis, a large South American scenario obtained from the Defense Planning Scenarios overstressed the system of existing facility locations, preventing the satisfaction of a 10-day IOC deadline, even with the use of APF ships. Although most states in South America are relatively stable, the recent difficulties in Ecuador and Venezuela demonstrate the potential volatility of the region. As with Africa, future U.S. intervention cannot be discounted, owing to significant U.S. interests in the region's oil supply. Although the current combat support infrastructure is sufficient for small-scale operations, such as drug interdiction, an expanded combat support presence would facilitate larger-scale operations in the region.

A multimodal transportation option is the key to rapid logistics response. RAND has shown in several earlier reports (Amouzegar et al., 2004, and Vick et al., 2002) that overreliance on airlift may in fact reduce response capability because of throughput constraints and availabity of airlift. A comprehensive mobility plan should include a combination of air, land, and sealift. Judicious use of trucks and high-speed sealift in fact may offer a faster and less expensive way to meet the Air Force's mobility needs.

Airlifter and Refueler Characteristics

One of the major factors in selecting a forward support location is its transport capability and capacity, and that capability and capacity can dictate the type of aircraft that can be used at a base and the load capacity it can handle. The tables in this appendix present the characteristics of various aircraft of interest.[1]

Table A.1
Aircraft Size

Aircraft	Length (feet)	Width (feet)[a]	Maximum Weight (feet)	Parking Spots (C-141 = 1)
C-130	99.50	132.60	175,000	0.50
C-141	168.40	160.00	343,000	1.00
C-17	173.92	169.75	585,000	1.13[b]
C-5A/B	247.80	222.70	840,000	2.00
KC-10	181.60	165.30	593,000	1.10
KC-135	136.25	130.85	322,500	0.70
B-747	231.83	195.67	836,000	1.70
DC-10	182.25	165.33	593,000	1.10

[a] Wingtip clearance: 10 feet on each side with wing walker, 25 feet on each side without wing walker. (The restrictions do not apply to the Civil Reserve Air Fleet.)
[b] With a wing walker, the C-17 can park in a C-141 spot.

---

[1] All data in this appendix are from U.S. Air Force (2003).

Table A.2
Aircraft Block Speeds

Aircraft	Mach	500 nmi	1,000 nmi	1,500 nmi	2,000 nmi	3,000 nmi	4,000 nmi	5,000 nmi
C-130	0.49	242	266	272	273	271	—	—
C-141	0.74	332	380	396	401	401	407	410
C-17	0.76	335	384	400	405	406	412	—
C-5A/B	0.77	341	393	410	415	416	422	426
KC-10	0.81	354	410	428	435	437	443	447
KC-135	0.79	348	401	419	426	430	435	438
B-747	0.84	363	422	442	450	452	459	463
DC-10	0.83	360	418	438	445	447	454	458

Table A.3
Aircraft Payloads

Aircraft	Pallet Position	Cargo (short tons) ACL[a]	Cargo (short tons) Planning[b]	Passengers ACL	Passengers Planning	NEO Passengers
C-130	6	17	12.0	90	80	92/74[c]
C-141	13	30	19.0	153	120	200/153 [c]
C-17	18	65	45.0	101	90	102
C-5A/B	36	89	61.3	73	51	73
KC-10[d]	23	60	32.6	75	68	75
KC-135	6	18	13.0	53	46	53
B-747	34	113	98.0	315	315	380
DC-10-1	30	79	69.0	242	242	350

NOTE: Cargo and passenger payload (except for the C-5) are exclusive of one another.
NE = noncombatant evacuation operation
[a] Organic cargo is calculated as the maximum allowable cabin load (ACL) for a 3,200-nmi leg; Civil Reserve Air Fleet (CRAF) is calculated for a 3,500-nmi leg.
[b] These numbers represent the historical average.
[c] The lower of the two numbers reflects life-raft capacity for noncombatant evacuation operations.
[d] Includes KC-10 (airlift) and KC-135 (airlift).

Table A.4
Ground Times

Aircraft	Passenger and Cargo Operations Wartime Planning Times (hours plus minutes)				Minimum Crew Rest Times (hours plus minutes)
	Onload	En Route	Offload	Expedited[b]	
C-130	2 + 15	1 + 30	2 + 15	0 + 45	16 + 15
C-141	3 + 15	2 + 15	3 + 15	1 + 15	16 + 00
C-17	3 + 15	2 + 15	3 + 15	1 + 45	16 + 30
C-5A/B	4 + 15	3 + 15	4 + 15	2 + 00	17 + 00
KC-10	4 + 15	3 + 15	4 + 15	3 + 15	17 + 00
KC-135	4 + 15	3 + 15	4 + 15	3 + 15	17 + 00
B-747	3 + 30/ 5 + 00[a]	1 + 30	3 + 30/ 5 + 00[a]	—	—
DC-10	2 + 30/ 5 + 00[a]	1 + 30	2 + 30/ 5 + 00[a]	—	—

[a] Includes passengers and cargo.

[b] Includes onload or offload operations only. Does not include refueling or reconfiguration operations.

Table A.5
Aircraft Utilization

Aircraft	UTE Rate[a]		Inventory[c]			
	Surge	Contingency (Sustained)	2004	2005	2006	2007
C-130	6.0	6.0	395	388	364	354
C-141	605	6.0	42	22	8	0
C-17	14.5	12.5	94	109	122	136
C-5A/B	8.5/11.5	7.7/8.1	96	94	94	94
KC-10[b,e]	9.8	8.6	54	54	54	54
KC-135[b,e]	6.8	5.1	445	429	421	421
CRAF[d]			Stage 1	Stage 2		Stage 3
B-747	10	10	11/6	27/20		85/61
DC-10	10	10	4/6	16/9		86/29

[a] The utilization rate (UTE rate) is the capability of a fleet of aircraft to generate flying hours in a day, expressed in terms of per Primary Authorized Inventory (PAI). Applies only to long-term, large-scale operations such as Operations Plans (OPLANs). For small operations involving less than the entire fleet, UTE rates are not normally a factor. Surge UTE rates apply for the first 45 days (C-130s surge for 30 days).

[b] KC-10 and KC-135 theater ballistic missile rates apply in the airlift role.

[c] Reflects active/Air Reserve Components aircraft inventory, not apportionment.

[d] CRAF cargo/passenger aircraft contracted for FY 2003.

[e] KC-135 and C-5 A/B primary mission aircraft inventory numbers are based on FY04 POM actions.

ROBOT User's Guide

This section presents a user's guide for the *RAND Overseas Basing Optimization Tool (ROBOT)*. The graphical user interface (GUI) allows the user to select the deployment scenarios and adjust all the pertinent parameters. The GUI is a link between a Microsoft Access database and ROBOT, which is written on a General Algebraic Modeling System (GAMS).[1]

Tutorial

We start this user's guide with a detailed example of the use of ROBOT for setting up and solving the problem of selecting a set of storage bases at the lowest possible cost given combat support requirements and other constrained resources.

In this example, the problem is to find the minimal-cost solution that meets the demands generated by two simultaneous deployments in Southeast Asia: a training operation in Singapore and a humanitarian relief operation in East Timor. In Singapore the operation will be conducted from Paya Lebar and in East Timor the Air Force has access to El Tari and Komodo.

In the opening screen (Figure B.1), accept the default values for the interest rate, hours per period, and percentage of munitions to retrograde. In addition, since the objective will be to minimize the

---

[1] For more information on GAMS and its solvers, see Brooke et al. (2003).

Figure B.1
Opening Screen of the ROBOT Tool

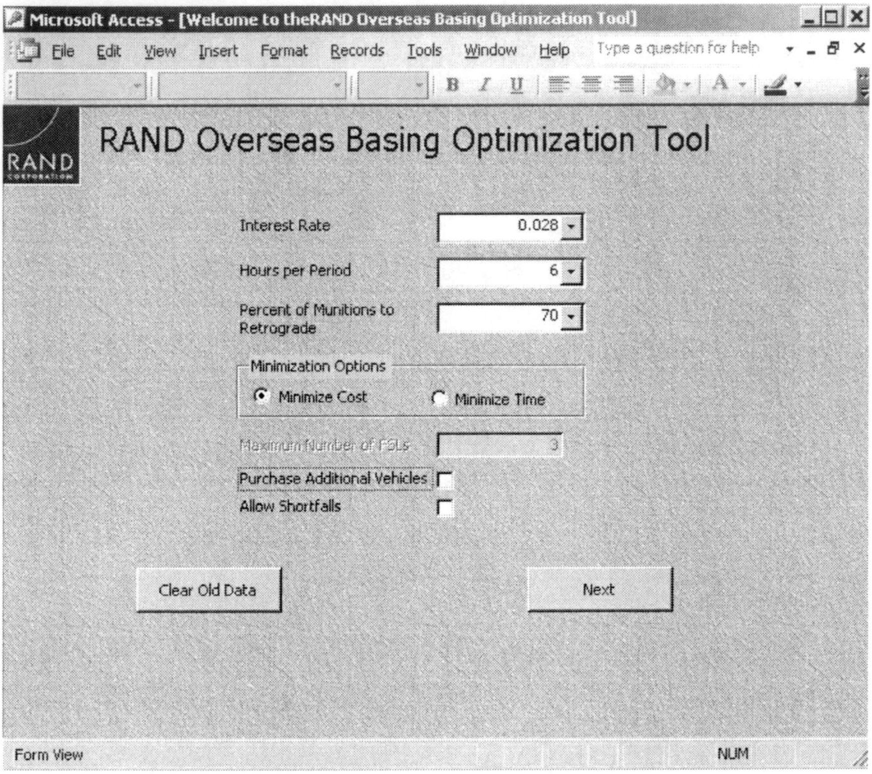

RAND *MG421-B.1*

cost needed to support these deployments, select the *minimize cost* option and turn off the checkbox for *Allow Shortfalls*. Then, click on *Next* at the bottom of the screen.

The next screen (Figure B.2) allows the user to set the contingencies that will make up the scenario. The example contingencies include one training operation and one humanitarian relief operation in Southeast Asia. Using the pull-down menu, we select *Singapore* and *East Timor*. Since we assume that these deployments occur simultaneously, the year column is filled with 1. Each contingency automatically generates a predetermined list of FOLs and the resource requirements at each FOL.

Figure B.2
Contingency Selection Tab

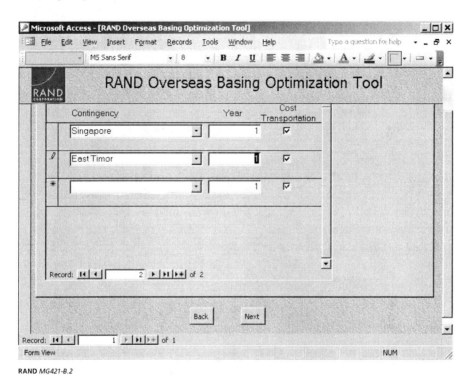

RAND *MG421-B.2*

The FSL Selection tab allows the user to select from a list of potential FSLs that might support the contingencies. In Figure B.3, there are five potential FSLs. Anderson AFB is *fixed on* (that is, it is selected a priori) because it is an existing bomber island with a large quantity of combat support resources. There are three potential land FSLs: Clark AB (Philippines), Darwin (Australia), and U-Tapao (Thailand). These locations are set as *variable* to allow the model to select them if needed. In addition, there is a potential munitions ship, *MUN2_Guam*, based in Guam. The throughput columns allow the user to set the MOG, land throughput, and seaport capability. For this tutorial, we use the default values. Click on *Next* at the bottom of the screen.

Figure B.3
FSL Selection

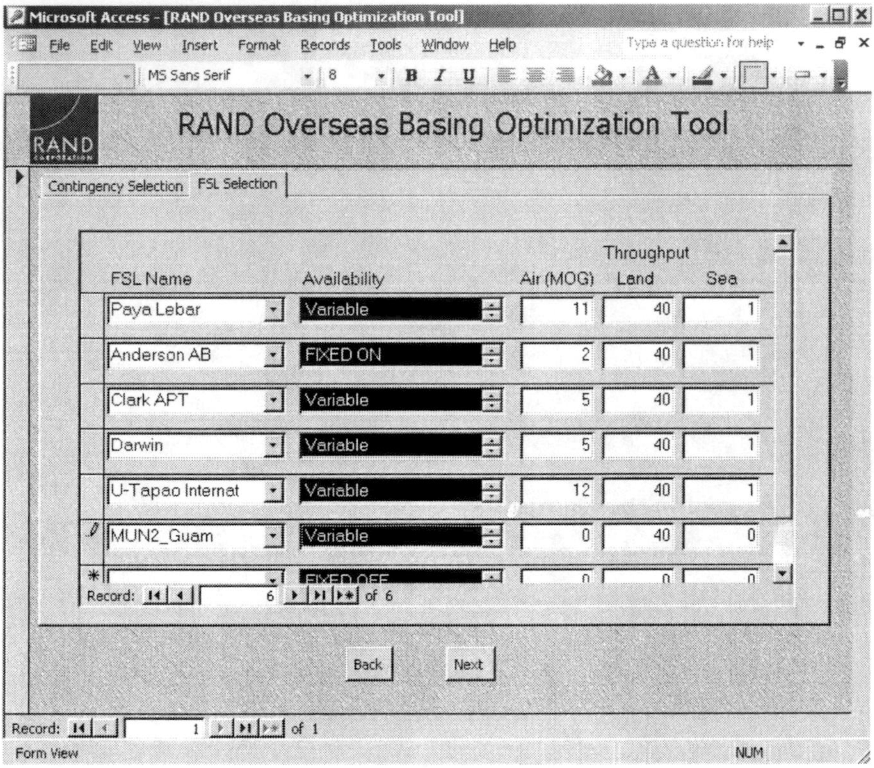

RAND *MG421-B.3*

The next step is to time-phase the combat support requirements and identify the characteristics of the forward operating locations (FOLs).

Figure B.4 illustrates the options for this example. The set of FOLs is determined automatically from the selected contingencies. In this case, Paya Lebar is opened to support training in Singapore and Komoro and El Tari are opened to support the humanitarian relief operation in East Timor. The user must define the time-phasing of the requirements by assigning values for the initial operating capability (IOC) date (in days), full operating capability (FOC) date,

Figure B.4
Scheduling the Requirements at the FOLs

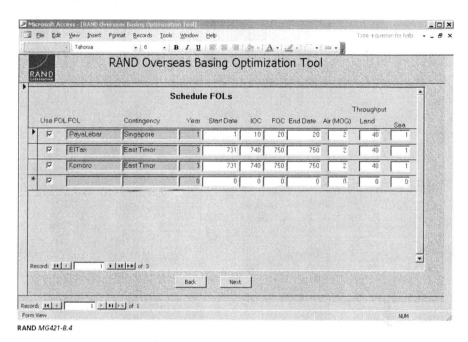

RAND *MG421-B.4*

the end of operations, and the throughput for each FOL. For this example, the IOC is set for 10 days and the FOC for 20 (not shown).

The MOG and other parameters are as indicated in Figure B.4. Click *Next* at the bottom of the screen.

The next step is to assign transport vehicles. All contingencies occurring in a given year draw from a common pool of transportation assets. Since in this case both contingencies occur in year 1, there is one entry for each vehicle type. For this tutorial, we assume that there are 10 C-17s, 15 C-130s, 2 high-speed sealift (HSS) vessels, and 100 trucks available to deliver resources from FSLs to the FOLs (see Figure B.5).

The next step is to build the model (i.e., create the data input for ROBOT and GAMS) and then solve the model. Clicking on the *Build and Solve* button creates an input file, and ROBOT is then run to solve the model. This process can take several minutes for a small problem and several hours for very large problems. There is also an

Figure B.5
Transportation Assets

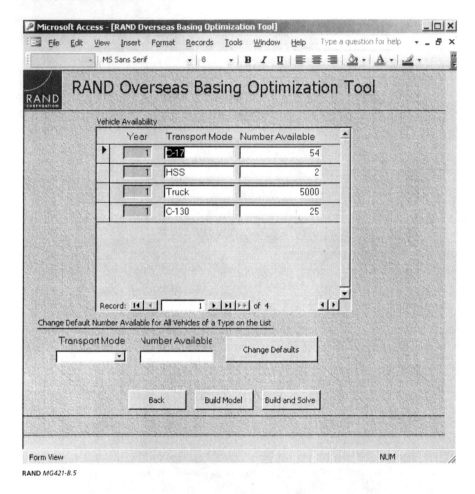

RAND MG421-B.5

option to create an input file only. This option allows the user to manipulate the input data directly before solving it. For this tutorial, however, click on the *Build and Solve* button.

The GUI will prompt for the location of the resulting output file as well as the location of the GAMS executable, if these locations have not been chosen before. In the *Please Select Directory* window that appears (see Figure B.6), select an output directory without

Figure B.6
Select Output Directory

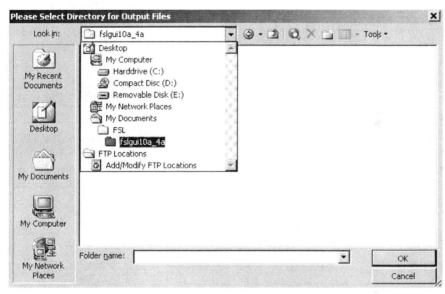

specifying a file name and click *OK*. Next, if the GUI needs to find the GAMS executable file (GAMS.EXE), another *Select Directory* window will appear. Select the directory that contains GAMS.EXE on your computer. The GUI will then create an input file as *online1.inc* and *online2.inc* in the chosen directory; if the *Build and Solve* option was chosen, the GUI will run GAMS/ROBOT. At the end of the run, an output file will be created with appropriate charts.

Since the *minimize cost* option was selected, the model will generate three types of costs—Transportation Cost, Facility Operating Cost, and Facility Construction Costs—as shown in Figure B.7. Had the *minimize time* (i.e. minimize time to IOC and FOC) objective been selected, the quantity of materiel (in tons) shipped from each FSL would be displayed, as shown in Figure B.8.

Figure B.7
Minimize Cost Output

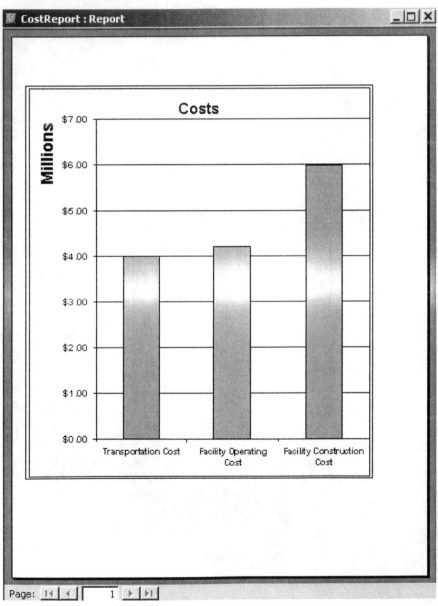

Figure B.8
Minimize Time Output

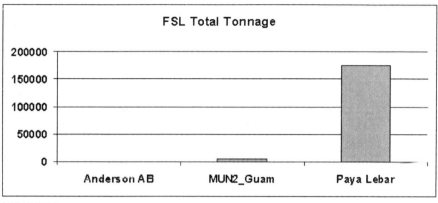

RAND *MG421-B.8*

The ROBOT User Interface

The user interface is designed to make the scheduling of contingencies, selection of candidate FSLs and vehicles, and the generation of the optimization input file straightforward. The various screens will set general parameters, schedule contingencies, select candidate FSLs, set IOC and FOC deadlines for the FOLs, and determine the number of available vehicles.

By basing the interface on a relational database, the GUI is able to utilize scenarios with a series of contingencies and candidate FSLs, over an arbitrary time period.

Opening Screen

In the opening screen (Figure B.9), the user has several options such as *interest rate*, the *hours per period*, *percent of munitions to retrograde*, *maximum number of FSLs*, the *objective* of the model (*minimize cost* or *minimize time*), and the options of *purchasing additional vehicles* and allowing a *shortfall*.

Figure B.9
ROBOT Opening Screen

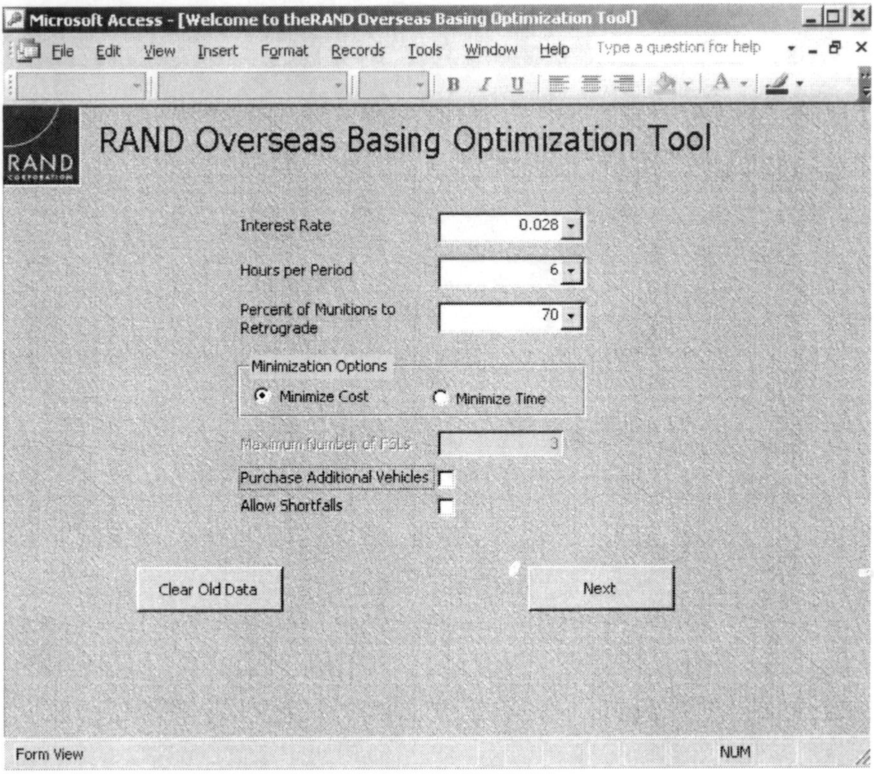

The *Interest Rate* box indicates the discount rate used to account for costs over time as the model allows the user to select several contingencies over a period of time. The default discount rate is set at 0.028, which is the Office of Management and Budget (OMB) rate for projects with a 10-year payback period (OMB, n.d.).

The *Hours per Period* box determines the resolution of the model. Within each contingency, time is divided into periods of the specified number of hours. In Figure B.9, the "6" in that box indicates that each day is divided into four six-hour periods. The duration of such activities as loading an aircraft or transit between two points is represented as a number of time periods. For example, suppose that it takes three hours to load a notional aircraft, and the transit time between two points via this aircraft is ten hours. Then, if a six-hour time period is used, this aircraft loading will require one time period and this transit will require two time periods. Using smaller time periods reduces the rounding error but requires more computational resources because of the increased modeling resolution.

The *Percentage of Munitions to Retrograde* box gives the percentage of munitions that are returned to the FSL (i.e., unused munitions) after the end of contingencies. The model costs for partial retrograde of munitions (due to expenditures), as well as for full retrograde of the base operating support and vehicles.

The *Minimization* options allow the user to choose between two options: minimizing the costs to support the contingencies, or minimizing the time needed to bring all FOLs to full operational capability. The *minimize cost* option determines the FSLs that lead to the least costs of construction, O&M, and peacetime transportation while also meeting the time-phased demand without violating the resource constraints. The *minimize time* option determines which FSLs would lead to the shortest time until closure of IOC/FOC, subject to a maximum allowable number of FSLs.

The *Allow Shortfalls* option relaxes the need for the system to meet IOC and FOC deadlines at each FOL for every contingency in the scenario. However, if this option is selected, there is an associated shortfall penalty imposed per period per ton delayed. The *Allow Shortfalls* option is automatically highlighted whenever the objective is to minimize time at the shortfall cost of one per ton per period, due to the characteristics of the mathematical model. The minimize time objective also requires that the *maximum allowable number of FSLs* be entered.[2]

The *Purchase Additional Vehicle* option allows for purchasing additional vehicles beyond the existing resources. The number of existing transport vehicles is entered in another screen.

---

[2] For details of the differences in the mathematical model between these two objectives, see Appendix C.

Contingency Selection

Figure B.10 illustrates the pull-down menu for *contingency selection.* Each contingency has several properties and requirements, including demand for munitions and vehicles, BEAR, the location of FOLs, the time requirements for IOC and FOC, and the access to FSLs. Some of these parameters are set a priori (e.g., munitions requirement) and can be changed only by editing the raw data; others require user input (e.g., time to IOC).

To select a contingency, simply click on the dropdown box as shown in Figure B.10 and select the year the deployment should start

Figure B.10
Contingency Selection

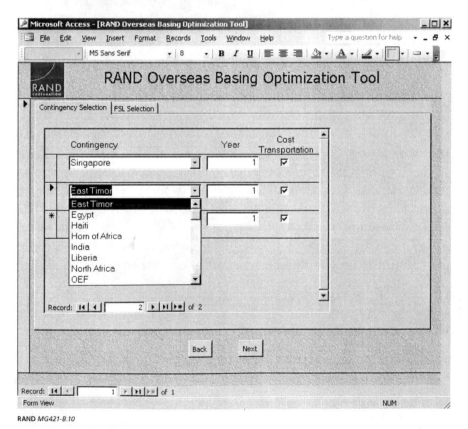

(e.g., year 5). In a later screen, the user will have an opportunity to schedule the contingency within the year. Multiple contingencies can occur in any given year. In addition, a contingency may be repeated in different years, as is the case with regularly scheduled exercises.

There is also an option to exclude the cost of transportation. As mentioned in Chapter Five, we assume that training exercises and other deployments intended to deter aggression are scheduled and all related costs including transportation should be included. However, major combat operations (MCOs) are funded using a different process; for the purpose of this study, MCO transportation cost is not factored in the optimization. This option allows the user to distinguish between different types of deployments.

FSL Selection

The model allows the user three options in the selection of the FSLs: force an FSL in the solution (*fixed on*), disallow an FSL to be considered (*fixed off*), or allow the model to select based on the optimization criteria (*variable*). Figure B.11 illustrates the various options available to the user.

The user also has the option of selecting the throughput capability of each FSL in terms of maximum-on-ground (MOG), land throughput and seaport capability. The MOG is measured in terms of the number of C-17s that can be handled at a given time. Land throughput is measured in terms of trucks per day. Sea throughput is measured in terms of the number of HSS vessels that can be served at a time. Although there is a default setting for each option, the user has the option of changing these parameters to account for changes in capability or to test the effect of mobility on the overall capability.

FOL Scheduling

After the contingencies are selected and scheduled, the model needs to account for the time-phased demand of materiel by assigning IOC

Figure B.11
FSL Selection

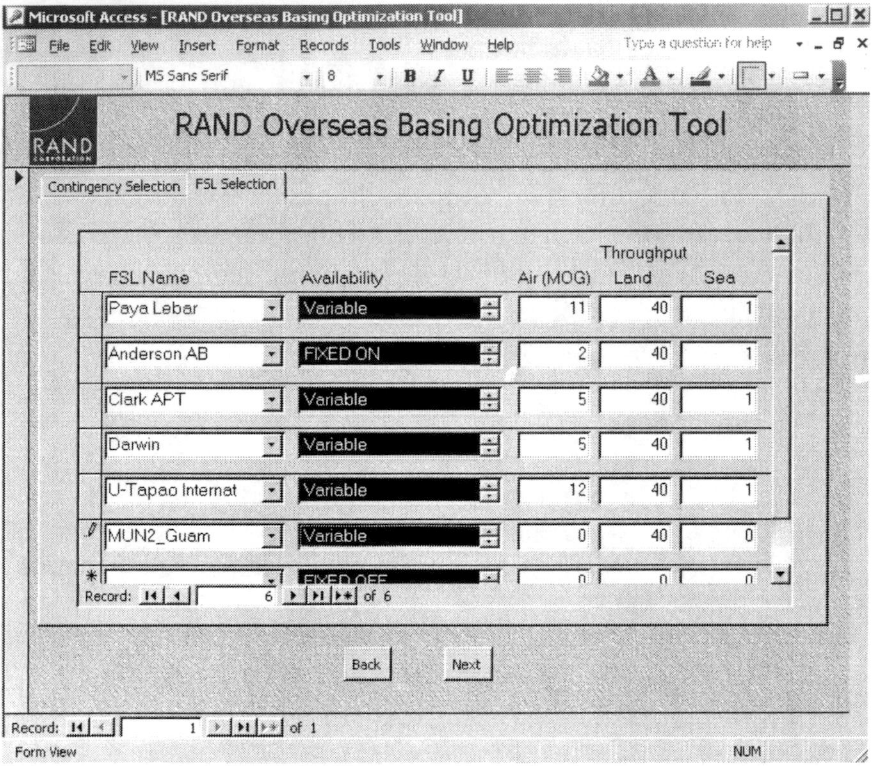

RAND MG421-B.11

and FOC dates for each of the FOLs. The default values for IOC and FOC dates are 10 days and 20 days, respectively, after the beginning of the year of the contingency, respectively. These may be changed to account for the urgency of the deployment or for more detailed scheduling considerations, using the FOL Scheduling screen, as illustrated in Figure B.12. For example, the Singapore contingency was selected to start in year 1 with the IOC Of 10 days. The OEF-like contingency is set for year 2 (e.g., start date of 366) with a 15-day IOC (day 380). Note that although a scenario can start at any date within a year, it must have an end date within its associated year.

Figure B.12
FOL Scheduling

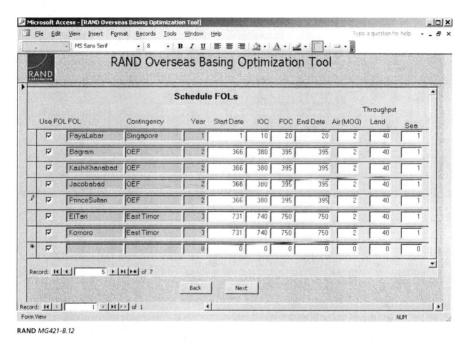

RAND MG421-B.12

In addition to the schedule for delivery of commodities, the throughput of the individual FOLs is determined. The default settings are MOG of two C-17 equivalents, 40 trucks per day, and one HSS. Note that when an FOL is inaccessible by sea transport, the default throughput of one HSS is irrelevant, as the model will prevent sea transportation from using the FOL. These values can be adjusted individually to account for changes in local conditions or for other use of the airfield, such as by other services or humanitarian relief efforts.

When the *Next* button is pressed, the model will pause for a few seconds as it updates several internal tables before displaying the *Vehicle Availability* screen.

Vehicle Selection

The *Vehicle Availability* screen indicates how many of each type of vehicle is available during each year that contingencies occur. There are currently four types of vehicles: C-17, C-130, High speed sealift (HSS), and Trucks. The parameters in Figure B.13 represent the numbers of each vehicle available for use for deployment in all contingencies in the given year.

Figure B.13
Vehicle Availability

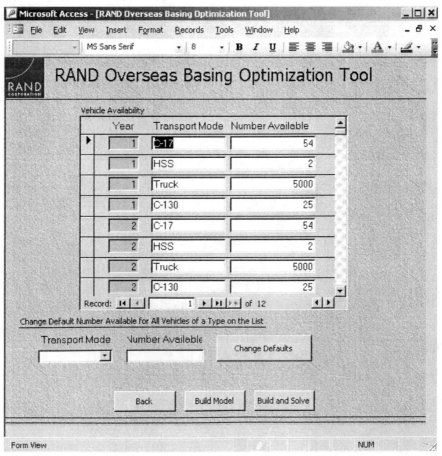

When the *Build Model* button is pressed, the model will present the option of choosing a directory for the output files as shown in Figure B.14. After the directory is chosen, the user should click on *OK* without indicating a file name. The model creates two files, *online1.inc* and *online2.inc* in the chosen directory. These data files are used as part of the optimization model.

After the determining the location of the output file directory, the model will ask for the location of the GAMS executable directory. Open the location of the GAMS directory and click *OK*.

Before explaining the optimization portion of the model, we show how the user can manipulate the *shortfall costs*.

Shortfall Costs

It is possible to relax the requirement to meet all demands on schedule by selecting the option on the first screen (Figure B.9). The

Figure B.14
Output Location Selection Window

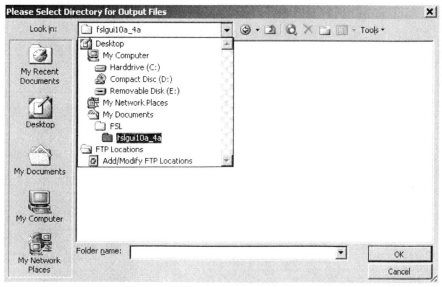

RAND *MG421-B.14*

model then relaxes the IOC and FOC requirements, allowing some materiel to be delivered later than planned with an added cost penalty. If this option is on, the *Edit Shortfall Costs* window (Figure B.15) will appear. This window allows the user to assign a shortage penalty for not delivering commodities by their IOC/FOC dates.

The shortfall cost values can be set to encourage or discourage late deliveries to the individual FOLs. The default value of 1,000,000 tons per day will discourage late deliveries. To allow some FOLs to have late deliveries, use smaller values for the shortfall penalty for individual commodities. It should be noted that when the option to *minimize time* is selected, these values would have a default setting of 1 ton per day (see Figure B.9). Because shortfall costs are the only cost components in the objective for *minimize time*, all IOC and FOC deadlines are set equal to the contingency start date, and this default shortfall cost is used to minimize the deployment time.

Figure B.15
Edit Shortfall Costs

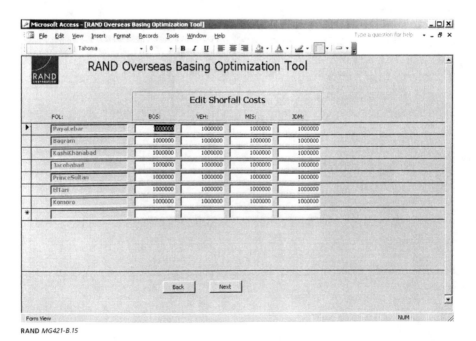

The Optimization Engine

As mentioned earlier, when the user clicks on the *Build and Solve* button, the model creates two files that are used as part of a GAMS/ROBOT input. The model takes candidate FSLs, resource availability and requirements, and the scheduled contingencies, and optimizes with respect to the selected objective (minimize cost or minimize time). The model then generates the following outputs:

- The optimal FSLs from the candidate locations
- The allocation of combat support resources to the selected FSLs
- The sizing of each storage location
- The transportation network connecting FSLs to FOLs, including the selection of the modes of the transport
- The cost of construction, transportation, and O&M.

Data Requirements

This section describes the data properties for some of the important parameters. This is important if the user needs to redefine the contingencies or update the characteristic of the locations or transport vehicles. All of these data are internal to the model and require a working knowledge of Microsoft Access and/or GAMS.

Forward Support Location

Each candidate FSL must have the following properties:

- Name
- Strip Name—GAMS uses an abbreviated form of the name of fewer than ten characters (no spaces) to identify the FSL in output
- Nearest port and port geo-code
- Distance to nearest port
- Geo-code—GAMS uses the geo-code internally to provide a unique identifier for the FSL location

- Speed factor—A land speed factor code to represent different speeds for developed (e.g., modern highway) to less developed (e.g. country roads)
- Cost factor—A cost multiplier for construction projects in a country. Also used for land transportation
- MOG—For air, based on C-17 equivalents; for sea, based on an HSS; for land, based on truck throughput per day
- Current number and properties of munitions igloos and warehouses on site
- Cost per additional square foot of munitions and nonmunitions storage space. In addition, operating costs for igloos and warehouses are set at $9.12 per square foot. Both the cost to build and operate igloos and warehouses need to be modified by the location cost factor.

These data are found in the Access table, *FSL Locations*. When this information is changed or a new FSL is added, other internal tables must be updated. This is done by running the Visual Basic function ImportFSLData().

Contingencies/FOL
Each contingency requires assets to be deployed to a number of FOLs in the region. A contingency is described by determining the quantity of assets that is required at each FOL. These are the characteristics of each FOL:

- Name
- Strip name
- Geo-code
- Nearest port and port geo-code
- Distance to nearest port
- Land speed factor
- Cost factor
- Quantity of Swift BEAR (SB) sets
- Quantity of 550i (Establish the base) sets
- Quantity of 550f (Operate the base) sets

- Quantity of Industrial Operations (IO) sets
- Quantity of Flight Line (FL) Sets
- Quantity of Follow-on Flight Line (FFL) sets
- Number of Rolling Stock (converted to short tons)
- Rolling Stock requirements in square feet
- Munitions (bombs)—short tons
- Munitions—pallets
- Net explosive weight (NEW)
- Missile—short tons
- Missile—pallets
- Missile NEW.

This information is stored in the table *START Output*. Once again the user must run the Visual Basic function ImportFOLData() to update other internal tables.

Each of the base operating sets (IO, FL, FFL, SB, 550i, 550f) is converted to tons transported and pallet positions required through the use of conversion factors based on START output (Snyder and Mills, 2004).

Vehicles

The transport vehicles are defined by their capacity (tons, square feet, and NEW for munitions cargo), loading and unloading time, passenger capacity, a space conversion factor, and a utilization rate. For aircraft, space is given as the space required compared to a C-17. For ships it is one berth for an HSS. For land vehicles the standard vehicle is a truck.

Airlift. Airlift information comes from AFPAM 10-1043 (U.S. Air Force, 2003) (see Appendix A). Air vehicles have a one-way cost. For the return trip, they assume a retrograde cost based on the lower of the air, sea, and land transit costs. Air distances are determined using the Joint Flow and Analysis System for Transportation (JFAST) (USTRANSCOM, 2002). From the air distances, transit time and costs are computed as well. For air transport, two tables include the *air cost* and the *air distance* for each FSL-FOL pair. Air cost is given as the cost per ton over the air link; distance is in nautical miles.

Sealift. The standard ship type is High Speed Sealift (HSS). The sustained speed is 30 knots with a capacity of 400 passengers or 400 tons of equipment. The transit time is based on JFAST port-to-port sailing distances of the ports associated with the FSL and FOL in question. Costs are based on MSC rates.[3] Retrograde for sea transport is assumed to be over sea links and carries the same cost. In addition, transit time for land transport is required in cases where the port associated with a given FSL or FOL is greater than 25 miles from the FSL/FOL. For sea transport, there are three tables: *distance, cost,* and *canal crossings.* Each FSL and FOL is assigned to a nearby port, if available. In addition to the sea transit time, when the port-to-FSL/FOL distance is more than 25 miles, land travel time is also added—based on the road conditions of the country that hosts the FOL. In addition to the distance and cost tables, there is a table that indicates the number of canal crossings required, referring to the Panama or the Suez canals. For each canal crossing required, a delay of one day's transit time is added to the total transportation time.

Land Transportation. A standard truck has a capacity of about 20 tons or 40 passengers. The distance required for land transport is based on output from JFAST. The transportation time is based on the road conditions of the country. For FSLs located in countries where the road conditions are "good," the daily distance covered by land transport is 194.4 miles per day. For FSLs located in countries where road conditions are "poor," the daily distance covered by land transport is 70.2 miles per day.[4] Cost for land transport is set at $0.25/(ton mile), adjusted for the country cost factor. For land transport, there are two tables: *distance* and *border crossings.* Each FSL-FOL pair has a land transport distance. The time for transit is determined by using the distance and the road conditions of the FSL. In addition, each border crossing results in a delay of one day (outside Europe).

---

[3] Rates are as shown in COMSCINST 7600.3J CH-10 MSC Billing Rates, October 4, 2002.

[4] Daily land transport speeds are taken from Headquarters, Department of the Army (1998).

Afloat Preposition Fleet (APF)

We allow for both munitions and nonmunitions APF ships. Each ship has a predetermined homeport and, for each given contingency, a port of call. We have assumed there are seven days of strategic warning prior to the contingency start day in which a ship can begin steaming toward its designated port.[5] After arriving at its port, the combat support equipment is delivered to FOLs using land transportation.

Pallet Positions

For vehicles, the area available is given in references as a number of pallet positions. Pallets positions are converted to area based on the size of a standard 463L pallet. Each pallet has a usable space of 84 × 104 inches (7 feet × 8 feet 8 inches) with a maximum build up to a height of 96 inches (8 feet) (U.S. Army Transportation School, 1997). Each vehicle type has its capacity expressed in terms of pallet positions. It is assumed that the entirety of the usable space of the pallet is used. Space requirements are determined in either square feet or in pallet positions. Requirements that are expressed in pallet positions are assumed to take up the entirety of their pallet positions.

Distance and Cost Tables

For each FSL-FOL pair, cost and distance tables are required for air, land, and sea modes of transport. These are stored in the *FSL-FOL tables*.

Materiel Storage

There are two types of storage at an FSL, igloos for munitions and warehouses for nonmunitions. Each igloo is assumed to have a storage space of 2,000 sq ft for 316,000 tons Net Explosive Weight (NEW) and costs $209/sq ft × the country cost factor to build an

---

[5] There is one exception. Because the South American scenario is not located near any potential FSLs, 19 days strategic warning for steam time is assumed for the that scenario.

additional igloo. Warehouses are 20,000 sq ft and cost $72/sq ft × the country cost factor to build. Operating costs for igloos and warehouses are $9.20/sq ft per year.[6]

---

[6] Construction costs are from Final FY 05 Pax Newsletter, http://www.hq.usace.army.mil/cemp/e/ec/ACF/321/FY05/Pax%203.2.1%20ACF%20CompleteR%20-%207Feb2005%20Final.pdf. Operating costs are based on $11.58/sq ft/yr at Sanem.

Mathematical Description of ROBOT

This appendix presents the mathematical programming formulation of the *RAND Overseas Basing Optimization Tool (ROBOT)*. For a detailed presentation of this model, including a descriptive explanation of the mathematics and a discussion of the model's limitations, see Amouzegar et al., 2004.

Sets and Set Indices

$i \in I$ commodities; $I = \{PAX, BOS, VEH, JDM, MIS, ...\}$

AM M(I)   munitions;

            AM M $(I) \subseteq I$; AM M$(I) = \{JDM, MIS, ...\}$

NAM(I)   nonmunitions;

            $NAM(I) \subseteq I$; $NAM(I) = \{BOS, VEH, ...\}$

$j \in J$ FSL index; $J = \{FSL1, FSL2, ..., APF1, APF2, ...\}$

AFL(J)   afloat FSLs;

            $AFL(J) \subseteq J$; $AFL(J) = \{APF1, APF2, ...\}$

$k \in K$ FOL index; $K = \{FOL1, FOL2, ...\}$

$m \in M$ modes of transport;
$$M = \left\{ C\text{-}130, C\text{-}17, C\text{-}5, B747, TRUCK, HSS, ... \right\}$$

$AIR(M)$ aircraft;
$$AIR(M) \subseteq M \; ; AIR(M) = \left\{ C\text{-}130, C\text{-}17, C\text{-}5, B747, ... \right\}$$

$LAN(M)$ land vehicles; $LAN(M) \subseteq M$;
$$LAN(M) = \left\{ TRUCK, ... \right\}$$

$SEA(M)$ sea vehicles; $SEA(M) \subseteq M$; $SEA(M) = \left\{ HSS, ... \right\}$

$PER(M)$ personnel transport vehicles; $PER(M) \subseteq M$;
$$PER(M) = \left\{ B747, HSS, ... \right\}$$

$h \in H$ phases; $H = \{1, 2, ...\}$

$o \in SCN$ deployment scenarios; $SCN = \left\{ 1, 2, ... \right\}$

$t \in T$ time periods that divide up each phase h; $T = \left\{ 1, 2, ... \right\}$

Data Parameters: Coefficients

Δ_j fixed cost incurred to open FSL j with $E_{\aleph j}$ square feet of storage space for commodity class $\aleph \left[AM\, M(I) = 1, NAM(I) = 2 \right]$

Θ_{mh} cost of obtaining an additional vehicle of mode m at the beginning of phase h

$\Xi_{\aleph j}$ construction cost per unit of storage needed beyond $E_{\aleph j}$ for commodity class $\aleph \lfloor AM\, M(I) = 1, NAM(I) = 2 \rfloor$ at FSL j

$\Upsilon_{\aleph j}$ operating cost (discounted over the time horizon) per unit of storage for commodity class $\aleph \lfloor AM\, M(I) = 1, NAM(I) = 2 \rfloor$ at FSL j

Ψ_{ik} shortfall cost per time unit per ton (or per passenger [PAX]) of commodity i not fulfilled at FOL k

Ω_{ijkm} cost per ton (or per PAX) of commodity i transported from FSL j to FOL k via mode m

α_m number of time periods necessary to load a mode m vehicle

β_m number of time periods necessary to unload a mode m vehicle

γ_m maximum load in tons per mode m vehicle

ε_m maximum load in square feet per mode m vehicle

ζ_k contingency start period at FOL k

η_k contingency finish period at FOL k

λ_m maximum load in PAX per mode m vehicle

μ_k phase of scenario associated with FOL k

π_{1jm} additional time needed prior to loading for commodities departing FSL j via mode m

π_{2km} additional time needed following unloading for commodities to reach FOL k via mode m

ρ_m conversion factor for parking space for mode m

σ_m utilization rate, expressed (for airlift) as the average flying hour goal per day divided by 24 hours, for mode m

τ_{jkm} one-way transportation time from FSL j to FOL k (or in opposite direction) via mode m

ϕ_{1i} conversion factor for commodity i from tons to square feet of storage space (= 0 for PAX)

ϕ_{2i} conversion factor for commodity i from tons to NEW

g_1 maximum square feet per munitions unit of storage

g_2 maximum square feet per nonmunitions unit of storage

g_3 maximum NEW per munitions unit of storage

$STEAM_{jk}$ additional time needed for steaming to port before afloat FSL j can begin offloading at port associated with FOL k

$FOLSCEN_k$ scenario associated with FOL k

$SCENPHASE_o$ phase associated with scenario o

Data Parameters

$A_{\aleph j}$ max on ground, in class $\aleph$
$$\left[AIR(M) = 1, LAN(M) = 2, SEA(M) = 3 \right] \text{ equivalent}$$
vehicles, at FSL j

$B_{\aleph k}$ max on ground, in class $\aleph$
$$\left[AIR(M) = 1, LAN(M) = 2, SEA(M) = 3 \right] \text{ equivalent}$$
vehicles, at FOL k

C_{mh} planned systemwide inventory of mode m vehicles at the beginning of phase h

D_{ikt} incremental demand, in tons (or PAX), for commodity i at FOL k at time t

$E_{\aleph j}$ minimum units of storage needed for an economically feasible FSL at location j for commodity class $\aleph$
$$\left[AMM(I) = 1, NAM(I) = 2 \right]$$

$F_{\aleph j}$ maximum potential units of storage at FSL j for commodity class $\aleph$ $\left[\text{AM M}(I) = 1, \text{NAM}(I) = 2 \right]$

Variables

$n_{\aleph j}$ additional units of storage needed beyond $E_{\aleph j}$ for commodity class $\aleph$ $\left[\text{AM M}(I) = 1, \text{NAM}(I) = 2 \right]$ at FSL j

p_{jkmt} number of mode m vehicles tasked to transport personnel from FSL j to FOL k, beginning loading on time t. **Integer**

q_{jmh} number of mode m vehicles available at FSL j at the start of time period $t = 1$ during phase h

r_{mh} additional mode m vehicles obtained at the beginning of phase h

s_{ikt} shortfall below demand, in tons (or PAX), for commodity i at FOL k not fulfilled by time t

u_{jkmt} number of nonsea mode m vehicles tasked to transport munitions from FSL j to FOL k, beginning loading on time t. **Integer**

v_{jmth} number of mode m vehicles available at FSL j at the end of time t during phase h

w_j **binary** variable indicating status of FSL j

x_{ijkmt} tons (or PAX) of commodity i sent from FSL j to FOL k via mode m, beginning loading at time t

y_{jkmt} number of nonsea mode m vehicles tasked to transport nonmunitions (or total sea mode m vehicles) from FSL j to FOL k, beginning loading at time t. **Integer**

z_{jkmt} number of mode m vehicles tasked to make the return trip from FOL k to FSL j, departing at time t. **Integer**

$ww_{\aleph j}$ units of storage utilized at FSL j for commodity class $\aleph \left[\text{AM M(I)} = 1, \text{NAM(I)} = 2 \right]$

pp_{jho} **binary** variable indicating if afloat FSL j supports scenario o during phase h

There is an implicit assumption throughout the entire model that terms having an index value $t \leq 0$ are not considered.

Objective Function

$$\min \sum_j \left(\begin{array}{c} \Delta_j w_j + \Xi_{"1"j} n_{"1"j} + \Xi_{"2"j} n_{"2"j} \\ + \Upsilon_{"1"j} ww_{"1"j} + \Upsilon_{"2"j} ww_{"2"j} \end{array} \right) \\ + \sum_{ijkmt} \Omega_{ijkm} x_{ijkmt} + \sum_{mh} \Theta_{mh} r_{mh} + \sum_{ikt} \Psi_{ik} s_{ikt} \tag{C.1}$$

Constraints

$$\sum_j q_{jmh} \leq \left(C_{mh} + r_{mh} \right) \qquad \forall m,h \tag{C.2}$$

$$\sum_{k \ni \mu_k = h} [p_{jkmt} + u_{jkmt} + y_{jkmt}] \leq v_{jm(t-1)h} \qquad \forall j,m,h; t \geq 2 \tag{C.3}$$

$$\sum_{k \ni \mu_k = h} \sum_{m \in \text{AIR(M)}} \sum_{n=0}^{\alpha_m - 1} [\rho_m (p_{jkm(t-n)} + u_{jkm(t-n)} + y_{jkm(t-n)})] \\ \leq A_{"1"j} \qquad \forall j,t,h \tag{C.4}$$

$$\sum_{k \ni \mu_k = h} \sum_{m \in LAN(M)} \sum_{n=0}^{\alpha_m - 1} [\rho_m (p_{jkm(t-n)} + u_{jkm(t-n)} + y_{jkm(t-n)})] \quad (C.5)$$

$$\leq A_{''2''j} \qquad \forall j, t, h$$

$$\sum_{k \ni \mu_k = h} \sum_{m \in SEA(M)} \sum_{n=0}^{\alpha_m - 1} [\rho_m (p_{jkm(t-n)} + u_{jkm(t-n)} + y_{jkm(t-n)})] \quad (C.6)$$

$$\leq A_{''3''j} \qquad \forall j, t, h$$

$$\sum_{j} \sum_{m \in AIR(M)} \sum_{n=0}^{\beta_m - 1} [\rho_m (p_{jkm(t-\tau_{jkm}-\alpha_m-n)} + u_{jkm(t-\tau_{jkm}-\alpha_m-n)})$$

$$+ y_{jkm(t-\tau_{jkm}-\alpha_m-n)}] \quad (C.7)$$

$$\leq B_{''1''k} \quad \forall k; \ \zeta_k \leq t \leq \eta_k$$

$$\sum_{j} \sum_{m \in LAN(M)} \sum_{n=0}^{\beta_m - 1} [\rho_m (p_{jkm(t-\tau_{jkm}-\alpha_m-n)} + u_{jkm(t-\tau_{jkm}-\alpha_m-n)})$$

$$+ y_{jkm(t-\tau_{jkm}-\alpha_m-n)}] \quad (C.8)$$

$$\leq B_{''2''k} \quad \forall k; \ \zeta_k \leq t \leq \eta_k$$

$$\sum_{j} \sum_{m \in SEA(M)} \sum_{n=0}^{\beta_m - 1} [\rho_m (p_{jkm(t-\tau_{jkm}-\alpha_m-n)} + u_{jkm(t-\tau_{jkm}-\alpha_m-n)})$$

$$+ y_{jkm(t-\tau_{jkm}-\alpha_m-n)}] \quad (C.9)$$

$$\leq B_{''3''k} \quad \forall k; \ \zeta_k \leq t \leq \eta_k$$

$$\sum_{j,m} \sum_{n=1}^{t} x_{ijkm(n-\tau_{jkm}-\alpha_m-\beta_m-\pi_{2km})}$$

$$\geq \left(\sum_{n=1}^{t} D_{ikn} \right) - s_{ikt} \qquad \forall i,k;\ \zeta_k \leq t \leq \eta_k \tag{C.10}$$

$$\sum_{k \ni \mu_k = h} \sum_{i \in \text{AMM(I)}} \sum_{m,t} \phi_{1i} x_{ijkmt} \leq g_1 \left(E_{''1''j} w_j + n_{''1''j} \right) \quad \forall j,h \tag{C.11}$$

$$\sum_{k \ni \mu_k = h} \sum_{i \in \text{AMM(I)}} \sum_{m,t} \phi_{2i} x_{ijkmt} \leq g_3 \left(E_{''1''j} w_j + n_{''1''j} \right) \quad \forall j,h \tag{C.12}$$

$$\sum_{k \ni \mu_k = h} \sum_{i \in \text{AMM(I)}} \sum_{m,t} \phi_{1i} x_{ijkmt} \leq g_1 w w_{''1''j} \quad \forall j,h \tag{C.13}$$

$$\sum_{k \ni \mu_k = h} \sum_{i \in \text{AMM(I)}} \sum_{m,t} \phi_{2i} x_{ijkmt} \leq g_3 w w_{''1''j} \quad \forall j,h \tag{C.14}$$

$$n_{''1''j} \leq \left(F_{''1''j} - E_{''1''j} \right) w_j \qquad \forall j \tag{C.15}$$

$$\sum_{k \ni \mu_k = h} \sum_{i \in \text{NAM(I)}} \sum_{m,t} \phi_{1i} x_{ijkmt} \leq g_2 \left(E_{''2''j} w_j + n_{''2''j} \right) \quad \forall j,h \tag{C.16}$$

$$\sum_{k \ni \mu_k = h} \sum_{i \in \text{NAM(I)}} \sum_{m,t} \phi_{1i} x_{ijkmt} \leq g_2 w w_{''2''j} \quad \forall j,h \tag{C.17}$$

$$n_{"2"j} \leq \left(F_{"2"j} - E_{"2"j}\right) w_j \qquad \forall j \tag{C.18}$$

$$\sum_{k,m,t} x_{"PAX"jkmt} \leq \left(\sum_{k,t} D_{"PAX"kt}\right) w_j \qquad \forall j \tag{C.19}$$

$$w_j \leq ww_{"1"j} + ww_{"2"j} \qquad \forall j \tag{C.20}$$

$$w_j \leq E_{"1"j} + n_{"1"j} + E_{"2"j} + n_{"2"j} \qquad \forall j \tag{C.21}$$

$$n_{"1"j} \leq ww_{"1"j} \qquad \forall j \tag{C.22}$$

$$n_{"2"j} \leq ww_{"2"j} \qquad \forall j \tag{C.23}$$

$$\sum_{k \ni \mu_k = h} \sum_j \left(\left[\sum_t \tau_{jkm} \left(p_{jkmt} + u_{jkmt} + y_{jkmt} \right) \right] + \left[\sum_{t=1}^{\|T\| - \tau_{jkm}} \tau_{jkm} z_{jkmt} \right] + \left[\sum_{t=\|T\| - \tau_{jkm} + 1}^{\|T\| - 1} \left(\|T\| - t \right) z_{jkmt} \right] \right)$$
$$\leq \|T\| \left(C_{mh} + r_{mh} \right) \sigma_m \quad \forall m, h \tag{C.24}$$

$$\sum_{i \in AMM(I)} x_{ijkmt} \le \gamma_m u_{jkmt} \quad \forall j,k; \; m \notin SEA(M);$$

$$\zeta_k + \pi_{1jm} \le t \le \eta_k - \tau_{jkm} - \alpha_m - \beta_m - \pi_{2km}$$

$$(C.25)$$

$$\sum_{i \in AMM(I)} \phi_{1i} x_{ijkmt} \le \varepsilon_m u_{jkmt} \quad \forall j,k; \; m \notin SEA(M);$$

$$\zeta_k + \pi_{1jm} \le t \le \eta_k - \tau_{jkm} - \alpha_m - \beta_m - \pi_{2km}$$

$$(C.26)$$

$$\sum_{i \in NAM(I)} x_{ijkmt} \le \gamma_m y_{jkmt}$$

$$\forall j,k; \; m \notin SEA(M);$$

$$\zeta_k + \pi_{1jm} \le t \le \eta_k - \tau_{jkm} - \alpha_m - \beta_m - \pi_{2km}$$

$$(C.27)$$

$$\sum_{i \in NAM(I)} \phi_{1i} x_{ijkmt} \le \varepsilon_m y_{jkmt} \quad \forall j,k; \; m \notin SEA(M);$$

$$\zeta_k + \pi_{1jm} \le t \le \eta_k - \tau_{jkm} - \alpha_m - \beta_m - \pi_{2km}$$

$$(C.28)$$

$$\sum_{i \in AMM(I) \cup NAM(I)} x_{ijkmt} \le \gamma_m y_{jkmt} \quad \forall j,k; \; m \in SEA(M);$$

$$\zeta_k + \pi_{1jm} \le t \le \eta_k - \tau_{jkm} - \alpha_m - \beta_m - \pi_{2km}$$

$$(C.29)$$

$$\sum_{i \in AMM(I) \cup NAM(I)} \phi_{1i} x_{ijkmt} \le \varepsilon_m y_{jkmt} \quad \forall j,k; \; m \in SEA(M);$$

$$\zeta_k + \pi_{1jm} \le t \le \eta_k - \tau_{jkm} - \alpha_m - \beta_m - \pi_{2km}$$

$$(C.30)$$

$$x_{"PAX"jkmt} \leq \lambda_m y_{jkmt} \qquad \forall j,k; \ m \in \left(\text{SEA(M)} \cap \text{PER(M)}\right);$$
$$\zeta_k \leq t \leq \eta_k$$

(C.31)

$$x_{"PAX"jkmt} \leq \lambda_m p_{jkmt} \qquad \forall j,k; \ m \in \text{PER(M)}; m \notin \text{SEA(M)};$$
$$\zeta_k \leq t \leq \eta_k$$

(C.32)

$$\sum_j z_{jkmt} = \sum_j [p_{jkm(t-\tau_{jkm}-\alpha_m-\beta_m)} + u_{jkm(t-\tau_{jkm}-\alpha_m-\beta_m)}$$
$$+ y_{jkm(t-\tau_{jkm}-\alpha_m-\beta_m)}] \forall k,m;$$
$$\zeta_k + \alpha_m + \beta_m \leq t \leq \eta_k - \pi_{2km}$$

(C.33)

$$v_{jmth} = v_{jm(t-1)h} + \sum_{k \ni \mu_k = h} [z_{jkm(t-\tau_{jkm})} - p_{jkmt} - u_{jkmt}$$
$$- y_{jkmt}] \quad \forall j,m,h; t \geq 2$$

(C.34)

$$v_{jm"1"h} = q_{jmh} + \sum_{k \ni \mu_k = h} [-p_{jkm"1"} - u_{jkm"1"} - y_{jkm"1"}]$$
$$\forall j,m,h$$

(C.35)

$$\sum_o pp_{jho} \leq 1 \qquad \forall h; j \in \text{AFL(J)}$$

(C.36)

$$\sum_{imt} \sum_{k \ni \text{FOLSCEN}_k = o} x_{ijkmt} \leq \left(\sum_{ikt} D_{ikt} \right)$$

$$\sum_{h=\text{SCENPHASE}_o} pp_{jho} \quad \forall o; \tag{C.37}$$

$$j \in \text{AFL(J)}$$

$$x_{ijkmt} = 0 \quad t < \zeta_k + \pi_{1jm}; \quad t > \eta_k - \tau_{jkm} - \alpha_m - \beta_m - \pi_{2km} \tag{C.38}$$

$$n_{"1"j}, n_{"2"j}, p_{jkmt}, q_{jmh}, r_{mh}, s_{ikt}, u_{jkmt}, v_{jmth},$$
$$x_{ijkmt}, y_{jkmt}, z_{jkmt}, ww_{"1"j}, ww_{"2"j} \geq 0 \tag{C.39}$$

$$p_{jkmt}, u_{jkmt}, y_{jkmt}, z_{jkmt} \quad \text{integer} \tag{C.40}$$

$$w_j, pp_{jho} \in \{0,1\} \tag{C.41}$$

Bibliography

Air Force Civil Engineer Support Agency, Directorate of Technical Support, *Historical Air Force Construction Cost Handbook*, Tyndall AFB, Fla.: HQ AFCESA, April 2000, http://www.afcesa.af.mil, last accessed August 2004.

Amouzegar, Mahyar A., Lionel A. Galway, and Amanda Geller, *Supporting Expeditionary Aerospace Forces: Alternatives for Jet Engine Intermediate Maintenance,* Santa Monica, Calif.: RAND Corporation, MR-1431-AF, 2002.

Amouzegar, Mahyar A., Robert S. Tripp, and Lionel A. Galway, "Integrated Logistics Planning for Expeditionary Aerospace Force," *Journal of Operational Research,* Vol. 55, 2004, pp. 422–430.

Amouzegar, Mahyar A., Robert S. Tripp, Ronald G. McGarvey, Edward W. Chan, and C. Robert Roll, Jr., *Supporting Air and Space Expeditionary Forces: Analysis of Combat Support Basing Options*, Santa Monica, Calif.: RAND Corporation, MG-261-AF, 2004.

Blanche, Ed, "Iran Forms Five Units for Shahab Ballistic Missiles," *Jane's Defence Weekly,* July 12, 2000.

Brooke, Anthony, David Kendrick, Alexander Meeraus, and Ramesh Raman, *General Algebraic Modeling System: A User's Guide,* Washington D.C.: GAMS Development Corporation, 2003.

Brunkow, Robert, and Kathyrn A. Wilcoxson, *Poised for the New Millennium: The Global Reach of the Air Mobility Command, A Chronology*, Scott Air Force Base, Ill.: Air Mobility Command, Office of History, April 2001.

Burger, Kim, "U.S. Moves Toward More Flexible Global Basing," *Jane's Defence Weekly,* May 2, 2003.

Byman, Daniel, Shahram Chubin, Anoushiravan Ehteshami, and Jerrold D. Green, *Iran's Security Policy in the Post-Revolution Era,* Santa Monica, Calif.: RAND Corporation, MR-1320-OSD, 2001.

Chubin, Shahram, "Whither Iran? Reform, Domestic Politics and National Security," *Adelphi Paper,* Issue 342, The International Institute for Strategic Studies, 2002.

Dipper, Martin, Jr., *91-Meter Wave Piercing Ferry INCAT 046, Transit from Hobart, Tasmania, Australia to Yarmouth, Nova Scotia, Canada,* Naval Surface Warfare Center, Carderock Division, CRDKNSWC/HD-1479-01, September 1998.

DoD—*see* U.S. Department of Defense

Eyal, Jonathan, "'Virtual' Military Bases on the Card for U.S.," *The Straits Times,* May 31, 2003.

Feinberg, Amatzia, Hyman L. Shulman, Louis W. Miller, and Robert S. Tripp, *Supporting Expeditionary Aerospace Forces: Expanded Analysis of LANTIRN Options,* Santa Monica, Calif.: RAND Corporation, MR-1225-AF, 2001.

Galway, Lionel A., Robert S. Tripp, Timothy L. Ramey, and John G. Drew, *Supporting Expeditionary Aerospace Forces: New Agile Combat Support Postures,* Santa Monica, Calif.: RAND Corporation, MR-1075-AF, 2000.

Galway, Lionel A., Mahyar A. Amouzegar, Richard J. Hillestad, and Don Snyder, *Reconfiguring Footprint to Speed Expeditionary Aerospace Forces Deployment,* Santa Monica, Calif.: RAND Corporation, MR-1625-AF, 2002.

Geller, Amanda, David George, Robert S. Tripp, Mahyar A. Amouzegar, and C. Robert Roll, Jr., *Supporting Air and Space Expeditionary Forces: Analysis of Maintenance Forward Support Location Operations,* Santa Monica, Calif.: RAND Corporation, MG-151-AF, 2004.

General Staff of the Bulgarian Army, *Military Strategy of the Republic of Bulgaria,* Sofia, Bulgaria: 2002, http://www.md.government. bg/_en_/docs/military_strategy.html, last accessed August 2004.

Global Security.org, "Sealift," 2005, http://www.globalsecurity.org/military/systems/ship/sealift.htm, last accessed October 10, 2005.

Groothuis, Todd, David Lyle, Robert Overstreet, Alan Lindsay, and John Dietz, *Afloat Prepositioning of Non-Munitions War Reserve Materiel (WRM) Phase II,* Maxwell AFB, Ala.: AFLMA Final Report LZ200125700, April 2003.

Headquarters, Department of the Army, *Armor and Mechanized-Based Opposing Force Operational Art,* Washington, D.C.: Field Manual 100-61, January 26, 1998.

House Armed Services Committee, "U.S. Forward Deployed Strategy in the European Theater," press release, February 26, 2003.

Institute for Strategic Studies (ISS), *The Military Balance 2002–2003,* London, UK: ISS, 2002.

Khalilzad, Zalmay, and Ian Lesser, *Sources of Conflict in the 21st Century: Regional Futures and U.S. Strategy,* Santa Monica, Calif.: RAND Corporation, MR-897-AF, 1999.

Killingsworth, Paul S., Lionel Galway, Eiichi Kamiya, Brian Nichiporuk, Timothy L. Ramey, Robert S. Tripp, and James C. Wendt, *Flexbasing: Achieving Global Presence for Expeditionary Aerospace Forces,* Santa Monica, Calif.: RAND Corporation, MR-1113-AF, 2000.

Lynch, Kristin F., John G. Drew, Robert S. Tripp, and C. Robert Roll, Jr., *Supporting Air and Space Expeditionary Forces: Lessons from Operation Iraqi Freedom,* Santa Monica, Calif.: RAND Corporation, MG-193-AF, 2005.

Office of Management and Budget, The Executive Office of the President, "Discount Rates for Cost Effectiveness, Lease Purchase, and Related Analyses," under "Real Discount Rates," OMB Circular No. A-94, n.d., http://www.whitehouse.gov/omb/circulars/a094/a94_appx-c.html, last accessed October 31, 2003.

Peltz, Eric, John Halliday, and Aimee Bower, *Speed and Power: Toward an Expeditionary Army,* Santa Monica, Calif.: RAND Corporation, MR-1755-A, 2003.

Peltz, Eric, Hyman L. Shulman, Robert S. Tripp, Timothy L. Ramey, and John G. Drew, *Supporting Expeditionary Aerospace Forces: An Analysis of F-15 Avionics Options,* Santa Monica, Calif.: RAND Corporation, MR-1174-AF, 2000.

Rainey, James C., Mahyar A. Amouzegar, Beth F. Scott, Robert S. Tripp, and Ann M. C. Gayer, eds., *Combat Support: Shaping Air Force Logistics for the 21st Century,* Maxwell AFB, Ala.: Air Force Logistics Management Agency, August 2003.

Shlapak, David A., John Stillion, Olga Oliker, and Tanya Charlick-Paley, *A Global Access Strategy for the U.S. Air Force,* Santa Monica, Calif.: RAND Corporation, MR-1216-AF, 2002.

Simon, Seena, "Bulgaria New Home to KC-135 Tanker Base," *Air Force Times,* August 6, 2003.

Snyder, Don, and Patrick H. Mills, *A Methodology for Determining Air Force Deployment Requirements,* Santa Monica, Calif.: RAND Corporation, MG-176-AF, 2004.

Snyder, Don, Patrick Mills, Manuel Carrillo, and Adam Resnick, *Supporting Air and Space Expeditionary Forces: Capabilities and Sustainability of Air and Space Expeditionary Forces Under Current and Alternative Policies,* Santa Monica, Calif.: RAND Corporation, MG-303-AF, 2005.

Sokolsky, Richard, Stuart E. Johnson, and F. Stephen Larrabee, *Persian Gulf Security: Improving Allied Military Contributions,* Santa Monica, Calif.: RAND Corporation, MR-1245-AF, 2000.

Stucker, James P., Ruth T. Berg, Andre Gerner, Amanda Giarla, William L. Spencer, Lory A. Arghavan, and Roy Gates, *Understanding Airfield Capacity for Airlift Operations,* Santa Monica, Calif.: RAND Corporation, MR-700-AF/OSD, 1998.

Sweetman, Bill, "Expeditionary USAF Sets Course," *Jane's International Defense Review,* Vol. 33, May 2000.

Tripp, Robert S., Lionel Galway, Paul S. Killingsworth, Eric Peltz, Timothy L. Ramey, and John G. Drew, *Supporting Expeditionary Aerospace Forces: An Integrated Strategic Agile Combat Support Planning Framework,* Santa Monica, Calif.: RAND Corporation, MR-1056-AF, 1999.

Tripp, Robert S., Lionel Galway, Timothy L. Ramey, Mahyar Amouzegar, and Eric Peltz, *A Concept for Evolving the Agile Combat Support/Mobility*

System of the Future, Santa Monica, Calif.: RAND Corporation, MR-1179-AF, 2000.

Tripp, Robert S., Kristin F. Lynch, John G. Drew, and Edward W. Chan, *Supporting Air and Space Expeditionary Forces: Lessons from Operation Enduring Freedom,* Santa Monica, Calif.: RAND Corporation, MR-1819-AF, 2004.

U.S. Air Force (USAF), "Air Force Basic Doctrine," *Air Force Doctrine Document 1* (AFDD-1), September 1, 1997.

————, *EAF Factsheet,* June 1999.

————, *Air Force Strategic Plan, Long-Range Planning Guidance,* Volume 3, Washington, D.C., May 2000.

————, *Air Mobility Planning Factors,* Washington D.C.: Air Force Pamphlet 10-1403, 2003.

————, Office of Force Transformation, *The Edge: Air Force Transformation,* April 2005, http://www.xp.hq.af.mil/xpx/docs/EDGEweb.pdf, last accessed January 2006.

U.S. Army Transportation School, *UMO Deployment Handbook, Reference 97-1, Appendix G,* Fort Eustice, Va.: U.S. Army Transportation School http://www.transchool.eustis.army.mil/UMOD/Guide/G.HTML, last accessed October 31, 2005.

U.S. Congressional Budget Study, *Options for Changing the Army's Overseas Basing,* Washington, D.C., May 2004.

U.S. Department of Defense (DoD), *Quadrennial Defense Review Report,* Washington, D.C.: September 30, 2001.

————, "U.S. Department of Defense Airlift Rates," August 2003a, http://public.amc.af.mil/fm/dodrates.doc, last accessed August 2003.

————, *Base Structure Report,* 2004, http://www.defenselink.mil/pubs/20040910_2004BaseStructureReport.pdf, last accessed October 5, 2005.

————, *Military Transformation: A Strategic Approach,* Fall 2003b, available for download at http://www.oft.osd.mil/index.cfm, last accessed October 28, 2005.

U.S. Navy Military Sealift Command, "Fact Sheet—Fast Combat Support Ships–T-AOE," August 2002, http://www.msc.navy.mil/factsheet/t-aoe.htm, last accessed August 2004.

———, "Fact Sheet—Fast Sealift Ships," December 2003, http://www.msc.navy.mil/factsheet/fss.htm, last accessed August 2004.

U.S. Transportation Command (USTRANSCOM), "Joint Flow and Analysis System for Transportation 8.0," USTRANSCOM, September 6, 2002, http://www.jfast.org/jfast/Manuals/jfast8/WebPageChapters%201-3.htm, last accessed October 28, 2005.

"U.S. to Realign Troops in Asia," *Los Angeles Times,* May 29, 2003.

Vick, Alan, David T. Orletsky, Bruce Pirnie, and Seth G. Jones, *The Stryker Brigade Combat Team, Rethinking Strategic Responsiveness and Assessing Deployment Options,* Santa Monica, Calif.: RAND Corporation, MR-1606-AF, 2002.